Green Beginnings: Joint Forest Management

Green Beginnings: Joint Forest Management

Ajay Mukherjee

RANDOM PUBLICATIONS
NEW DELHI (INDIA)

Green Beginnings: Joint Forest Management

ISBN 978-93-5111-834-3

Published in 2016 in India by

RANDOM PUBLICATIONS

4376-A/4B, Gali Murari Lal, Ansari Road
New Delhi-110 002
Phone : +9111-43580356, 011-23289044, 011-43142548
e-mail: sales@randompublications.com, info@randompublications.com, randomexports@gmail.com

Type Setting by : Friends Media, Delhi-110089
Printed at : Sanat Printers

Preface

Joint Forest Management (JFM) is now a principal forest management strategy in India. The government views JFM as a pivotal strategy for addressing the national policy goal of achieving 33% forest cover by 2012. JFM represents one model of community based forestry in which the state engages with communities with forestry. Though the current JFM model is weighted in favour of State Forest Department control over planning, management, Investment, harvesting and marketing.

Joint Forest Management often abbreviated as JFM is the official and popular term in India for partnerships in forest movement involving both the state forest departments and local communities. The policies and objectives of Joint Forest Movement are detailed in the Indian comprehensive National Forest Policy of 1988 and the Joint Forest Movement Guidelines of 1990 of the Government of India.

Originating in the early 1970s, the concept of joint forest management (JFM) is the official and popular term in India and elsewhere for partnerships in forest management involving both the state forest departments and local communities. Although schemes vary from state to state, the system works with villagers agreeing to assist in the safeguarding of forest resources through protection from fire, grazing and illegal harvesting in exchange for non-timber forest products and a share of the revenue from the sale of timber products. It was born in response to the many conflicts over forests, notably the Chipko movement of the 1970s in the Himalaya.

The book provides a vivid explanation to the forest resource and their importance in providing the livelihood to growing population and the need for sustainable forest management system.

– Author

Contents

1

Evergreen Forest

An evergreen forest is a forest consisting entirely or mainly of evergreen trees that retain green foliage all year round. Such forests reign in the equatorial region, between the tropics primarily as broadleaf evergreens, and in temperate and boreallatitudes primarily as coniferous evergreens.

Moist forest, montane forest,mossy forests, laurel forest, cloud forest, fog forest, are generally tropical or subtropical or mild temperate evergreen forest, found in areas with high humidity and relatively stable and mild temperatures, characterized by a persistent, frequent or seasonal low-level cloud cover, usually at the canopy level. Cloud forests often exhibit an abundance of mosses covering the ground and vegetation.

TROPICAL EVERGREEN FORESTS

Tropical evergreen forests are usually found in areas receiving more than 200 cm of rainfall and having a temperature of 15 °C to 30 °C and have annual humidity exceeding 77%. They occupy about seven per cent of the Earth's land surface and harbour more than half of the planet's terrestrial plants and animals. Tropical evergreen forests are dense, multi-layered, and harbour many types of plants and animals. The trees are evergreen as there is no period of drought or frost. The canopy tree species are mostly tall hardwoods with broad leaves that release large quantities of water through transpiration, in a cycle that is important in raising as much mineral nutrient material as possible from the soil.

In India, evergreen forests are found on the eastern and western slopes of the Western Ghats in such states as Tamil Nadu, Karnataka, Kerala and Maharashtra.

And also found in Assam, Arunachal Pradesh, Meghalaya, Nagaland, Tripura, West Bengal and Andaman and Nicobar Islands. They are also found in the hills of Jaintia and Khasi. Some of the trees found in Indian tropical forests are rosewood, mahogany and ebony. Bamboo and reeds are also common. Because of dense foliage competing for light, little direct sunlight reaches the understory.

TEMPERATE EVERGREEN FORESTS

Temperate evergreen forests, coniferous, broadleaf, and mixed, are found largely in the temperate mid-latitudes of Montane North America, Siberia, Canada, Australia, Africa and Scandinavia. Broadleaf evergreen forests occur in particular in eastern North America and in countries around the Mediterranean Basin, such as Lebanon and Morocco. Many subtropical broadleaved evergreen forests occur along the eastern margins of major land masses, e.g., in southeastern United States, southern China and in southeastern Brazil. Other examples include the wet temperate conifer forests of northwestern North America. Temperate evergreen forests are the regional climax vegetation, commonly dominated by hardy trees that can deal with sandy, rocky, and various other soils of poor quality. Most such communities also are subject to intermittent fire, drought and cold.

LAUREL FOREST

Fig. Laurel forest in Madeira

Fig. Laurel forest in Tenerife (Anaga)

Coniferous temperate evergreen forests are most frequently dominated by species in the familiesPinaceae and Cupressaceae. Broadleaf temperate evergreen forests include those in which Fagaceae such as oaks are common, those in which Nothofagaceae predominate, and the Eucalyptus forests of the Southern Hemisphere. There also are assorted temperate evergreen forests dominated by other families of trees, such as Lauraceae in laurel forest. Evergreen forests around the world are under threat of logging, mining, oil and gas developments, pollution, hydroelectric projects and other human developments planned in these areas.

Laurel forest, also called laurisilva or laurissilva, is a type of subtropical forest found in areas with high humidity and relatively stable, mild temperatures. The forest is characterized by broadleaf tree species with evergreen, glossy and elongated leaves, known as "laurophyll" or "lauroid". Plants from the laurel family (Lauraceae) may or may not be present, depending on the location.

SETTING

Fig. Laurel rain forest in La Gomera

Laurel forests are specific to wet forests from sea level to the highest mountains, but are poorly represented in areas with a pronounced dry season. They need an ecosystem of high humidity, such as cloud forests, with abundant rainfall throughout the year.

Laurel forests typically occur on the slopes of tropical or subtropical mountains, where the moisture from the ocean condenses so that it falls as rain or fog and soils have high moisture levels.

Some evergreen tree species will survive short frosts, but most species will not survive hard freezes and prolonged cool weather. They need a mild climate with annual temperature oscillation moderated by the proximity of the ocean. These conditions of temperature and moisture occur in four different geographical regions:

- Along the eastern margin of continents at latitudes of 25° to 35°.
- Along the western continental coasts between 40° and 55° latitude.
- On islands between 25° and 35° or 40° latitude.
- In humid montane regions of the tropics.

Some laurel forests are a type of cloud forest. Cloud forests are found on mountain slopes where the dense moisture from the sea or ocean is precipitated as warm moist air masses blowing off the ocean are forced upwards by the terrain, which cools the air mass to the dew point. The moisture in the air condenses as rain or fog, creating a habitat characterized by cool, moist conditions in the air and soil. The resulting climate is wet and mild, with the annual oscillation of the temperature moderated by the proximity of the ocean.

CHARACTERISTICS

Laurel forests are characterized by evergreen and hardwood trees, reaching up to 40 metres (130 ft) in height. Laurel forest, laurisilva, and laurissilva all refer to plant communities that resemble the laurel bay.

Some species belong to the true laurel family or Lauraceae, but many have similar foliage to the Lauraceae due to convergent evolution. As in any other rainforest, plants of the laurel forests must adapt to high rainfall and humidity. The trees adapted in response to these ecological drivers by developing analogous structures, leaves that repel water.Laurophyll or lauroid leaves are characterized by a generous layer of wax, making them glossy in appearance, and a narrow, pointed oval shape with an *apical mucro* or "drip tip", which permits the leaves to shed water despite the humidity, allowing perspiration and respiration. The scientific names *laurina*, *laurifolia*, *laurophylla*, *lauriformis*, and*lauroides* are often used to name species of other plant families that resemble the Lauraceae. The term *Lucidophyll*, referring to the shiny surface of the leaves, was proposed in 1969 by Tatuo Kira. The scientific names *Daphnidium*, *Daphniphyllum*, *Daphnopsis*, *Daphnandra*, *Daphne* from Greek: ÄÜöíç, meaning "laurel", *laural*, *Laureliopsis*,*laureola*, *laurelin*, *Laurelindorinan*, *laurifolia*, *Cistus laurifolius* (Laurel Rockrose), *laurifolius*, *lauriformis*, *laurina*, *laurophylla*, *laurocerasus*, *laurus*, *Prunus laurocerasus* (English laurel), *Prunus lusitanica* (Portugal laurel), *Corynocarpus laevigatus* (New Zealand Laurel), and *Corynocarpus rupestris* designate species of other plant families that resemble Lauraceae. The term "lauroid" is also applied to climbing plants such as ivies whose leaves resemble those of the Lauraceae.

Mature laurel forests typically have a dense tree canopy and low light levels at the forest floor. Some forests are characterized by an overstory of emergent trees.

Laurel forests are typically multi-species, and diverse in both the number of species and the genera and families represented. In the absence of strong environmental selective pressure, the number of species sharing the arboreal stratum is high, although not reaching the value of tropical forests; nearly 100 tree species have been described in the laurisilva rainforest of Misiones (Argentina), about 20 in the Canary Islands. This species diversity contrasts with other temperate forest types, which typically have a canopy dominated by one or a few species. Species diversity generally increases towards the tropics. In this sense, the laurel forest is a transitional type between temperate forests and tropical rainforests.

ORIGIN

Laurel forests are composed of vascular plants that evolved millions of years ago. Lauroid floras have included forests of Podocarpaceae and southern beech.

This type of vegetation characterized parts of the ancient supercontinent of Gondwana and once covered much of the tropics. Some lauroid species that are found outside laurel forests are relicts of vegetation that covered much of the mainland of Australia, Europe, South America, Antarctica, Africa, and North America when their climate was warmer and more humid. Cloud forests are believed to have retreated and advanced during successive geological eras, and their species adapted to warm and wet conditions were replaced by more cold-tolerant or drought-tolerant sclerophyll plant communities. Many of the late Cretaceous - early Tertiary Gondwanan species of flora became extinct, but some survived as relict species in the milder, moister climate of coastal areas and on islands. Thus Tasmania and New Caledonia share related species extinct on the Australian mainland, and the same case occurs on the Macaronesia islands of the Atlantic and on the Taiwan, Hainan, Jeju, Shikoku, Kyûshû, and Ryûkyû Islands of the Pacific.

Although some remnants of archaic flora, including species and genera extinct in the rest of the world, have persisted as endemic in such coastal mountain and shelter sites, their biodiversity was reduced. Isolation in these fragmented habitats, particularly on islands, has led to the development of vicariant species and genera. Thus, fossils dating from before the Pleistocene glaciations show that species of *Laurus* were formerly distributed more widely around the Mediterranean and North Africa. Isolation gave rise to*Laurus azorica* in the Azores Islands, *Laurus nobilis* on the mainland, and *Laurus novocanariensis* in the Canary Islands.

LAUREL FOREST ECOREGIONS

Laurel forests occur in small areas where their particular climatic requirements prevail, in both the northern and southern hemispheres. Inner laurel forest ecoregions, a related and distinct community of vascular plants, evolved millions of years ago on the supercontinent of Gondwana, and species of this community are now found in several separate areas of the Southern Hemisphere, including southern South America, southernmost Africa, New Zealand, Australia and New Caledonia. These lauroid plant communities have a twofold origin in tropical climates and temperate climates, those in temperate climates having their origin in the Antarctic flora.

Most Laurel forest species are evergreen, and occur in tropical, subtropical, and mild temperate regions and cloud forests of the northern and southern hemispheres, in particular the Macaronesian islands, southern Japan, Madagascar, New Caledonia, Tasmania, and central Chile, but they are pantropical, and for example in Africa they are endemic in the Congo region, Cameroon, Sudan, Tanzania, and Uganda, in lowland forest and Afromontane areas. Since laurel forests are archaic populations that diversified as a result of isolation on islands and tropical mountains, their presence is a key to dating climatic history.

EAST ASIA

Laurel forests are common in warm-temperate to subtropical eastern Asia, and form the climax vegetation in southern Japan, southern Korea, Taiwan, southern China, the mountains of Indochina, and the eastern Himalayas. In southern China, laurel forest once extended throughout the Yangtze Valley and Sichuan Basin from the East China Seato the Tibetan Plateau. The northernmost laurel forests in East Asia occur at 39° N. on the Pacific coast of Japan. Altitudinally, the forests range from sea-level up to 1000 metres in warm-temperate Japan, and up to 3000 metres elevation in the tropical mountains of Asia.

Some forests are dominated by Lauraceae, while in others evergreen laurophyll trees of the beech family (Fagaceae) are predominant, including ring-cupped oaks (*Quercus* subgenus *Cyclobalanopsis*), chinquapin (*Castanopsis*) and tanoak (*Lithocarpus*). Other characteristic plants include *Schima* and *Camellia*, which are members of the tea family (Theaceae), as well as magnolias, bamboo, andrhododendrons. These subtropical forests lie between the temperate deciduous and conifer forests to the north and the tropical monsoon forests of Indochina and India to the south.

Associations of Lauraceous species are common in broadleaved forests; for example, *Litsea* spp., *Litsea cupola, Persea odoratissima, Persea duthiei,* etc., along with such others as *Engelhardtia spicata, Rhododendron arboreum, Lyonia ovalifolia, Pyrus pashia, Rhus* spp., *Acer oblongum, Myrica esculenta, Michelia kisopa*, and *Betula alnoides*. Some other common trees and large shrub species of subtropical forests are *Semicarpus anacardium, Cretaeava unilocularis, Trewia nudiflora, Premna interrupta, Ulmus lancifolia, Ulmus chumlia, Glochidium velutinum, Callicarpa arborea, Toona ciliata, Ficus* spp., *Mahosama similicifolia, Trevesia palmate, Xylosma longifolium, Boehmeria rugulosa, Scheffera venulosa, Michelia* spp., *Casearia graveilens, Rhus wallichii, Actinodaphne reticulata, Sapimum insegne, Alnus nepalensis, Ardisia thyrsiflora, Ilex* spp,*Macaranga pustulata, Trichilia cannoroides, Celtis tetranda, Wenlendia puberula, Saurauia nepalensis, Ligustrum confusum, Quercus glauca, Zizyphus incurva, Camellia kissi, Hymenodictyon flaccidum, Maytenus thomsonii, Zanthoxylum armatum,*

Rhus succednea, Eurya acuminata, Myrsine semiserrata, Slonea tomentosa, Hydrangea asper, Symplocus spp., *Cleyrea* spp. and *Quercus lamellose*.

In the temperate zone, the cloud forest between 2,000 and 3,000 m altitude supports broadleaved evergreen forest dominated by plants such as *Quercus lamillosa* and *Q. semicarpifolia* in pure or mixed stands. Species such as *Lindera* and *Litsea*, *Tsuga dumosa*, and *Rhododendron* spp. are also present in the upper levels of this zone. Other important species are *Magnolia campbellii, Michelia doltsopa, Pieris ovalifolia, Daphnephyllum himalayanse, Acer campbellii, Acer pectinatum*, and *Sorbus cuspidata*, but these species do not extend toward the west beyond central Nepal. *Alnus nepalensis*, a pioneer tree species, grows gregariously and forms pure patches of forests on newly exposed slopes, in gullies, beside rivers, and in other moist places.

The common forest types of this zone include *Rhododendron arboreum, Rhododendron barbatum, Lyonia* spp., *Pieris formosa; Tsuga dumosa* forest with such deciduous species as *Acer* and *Magnolia*; deciduous mixed broadleaved forest of *Acer campbellii, Acer pectinatum, Sorbus cuspidata*, and *Magnolia campbellii*; mixed broadleaved forest of *Rhododendron arboreum, Acer campbellii, Symplocos ramosissima* and Lauraceae.

This zone is habitat for many other important tree and large shrub species such as *Abies pindrow, Betula utilis, Buxus rugulosa, Benthamidia capitata, Corylus ferox, Deutzia staminea, Euonymus tingens, Abies spectalbilis, Acanthopanax cissifolius, Acer campbellii, Acer pectinatum, Betula alnoides, Coriaria terminalis, Fraxinus macrantha, Dodecadenia grandiflora, Eurya cerasifolia, Hydrangea heteromala, Ilex dipyrena, Ligustrum* spp., *Litsea elongata, Juglans regia, Lichelia doltsopa, Myrsine capitallata, Neolitsea umbrosa, Philadelphus tomentosus, Osmanthus fragrans, Prunus cornuta, Rhododendron companulatum, Sorbus cuspidate*, and *Viburnum continifolium*.

In ancient times, laurel forests (*shoyojurin*) were the predominant vegetation type in the Taiheiyo evergreen forests ecoregion of Japan, which encompasses the mild temperate climate region of southeastern Japan's Pacific coast. There were three main types of evergreen broadleaf forests, in which *Castanopsis*, *Machilus*, or *Quercus* predominated. Most of these forests were logged or cleared for cultivation and replanted with faster-growing conifers, like pine or hinoki, and only a few pockets remain.

LAUREL FOREST ECOREGIONS IN EAST ASIA

- Changjiang Plain evergreen forests (China)
- Chin Hills-Arakan Yoma montane rain forests (Myanmar)
- Eastern Himalayan broadleaf forests (Bhutan, India, Nepal)
- Guizhou Plateau broadleaf and mixed forests (China)
- Jiang Nan subtropical evergreen forests (China)
- Nihonkai evergreen forests (Japan)

- Northern Annamites rain forests (Laos, Vietnam)
- Northern Indochina subtropical forests (China, Laos, Myanmar, Thailand, Vietnam)
- Northern Triangle subtropical forests (Myanmar)
- Sichuan Basin evergreen broadleaf forests (China)
- South China-Vietnam subtropical evergreen forests (China, Vietnam)
- Southern Korea evergreen forests (South Korea)
- Taiheiyo evergreen forests (Japan)
- Taiwan subtropical evergreen forests (Taiwan)
- Yunnan Plateau subtropical evergreen forests (China)

MALAYSIA, INDONESIA, AND THE PHILIPPINES

Laurel forests occupy the humid tropical highlands of the Malay Peninsula, Greater Sunda Islands, and Philippines above 1000 metres elevation. The flora of these forests is similar to that of the warm-temperate and subtropical laurel forests of East Asia, including oaks *(Quercus)*, tanoak *(Lithocarpus)*, chinquapin *(Castanopsis)*, Lauraceae,Theaceae, and Clethraceae. Epiphytes, including orchids, ferns, moss, lichen, and liverworts, are more abundant than in either temperate laurel forests or the adjacent lowland tropical rain forests.

Myrtaceae are common at lower elevations, and conifers and rhododendrons at higher elevations. These forests are distinct in species composition from the lowland tropical forests, which are dominated by Dipterocarps and other tropical species.

LAUREL FOREST ECOREGIONS OF SUNDALAND, WALLACEA, AND THE PHILIPPINES

- Borneo montane rain forests
- Eastern Java-Bali montane rain forests
- Luzon montane rain forests
- Mindanao montane rain forests
- Peninsular Malaysian montane rain forests
- Sulawesi montane rain forests
- Sumatran montane rain forests
- Western Java montane rain forests

MACARONESIA AND THE MEDITERRANEAN BASIN

Laurel forests are found in the islands of Macaronesia in the eastern Atlantic, in particular the Azores, Madeira Islands, and Canary Islands from 400 to 1200 metres elevation. Trees of the genera *Apollonias* (Lauraceae), *Ocotea* (Lauraceae), *Persea* (Lauraceae),*Clethra* (Clethraceae), *Dracaena* (Ruscaceae), and *Picconia* (Oleaceae) are characteristic. The Madeira Islands laurel forest was designated a World Heritage Site by UNESCO in 1999.

Millions of years ago, laurel forests were widespread around the Mediterranean Basin. The drying of the region since the Plioceneand cooling during the Ice Ages caused these rainforests to retreat. Some relict Mediterranean laurel forest species, such as Sweet Bay *(Laurus nobilis)* and European Holly *(Ilex aquifolium)*, are fairly widespread around the Mediterranean basin.

In the Mediterranean there are other areas with species adapted to the same habitat, but which do not form a laurel forest. The most important is the ivy, a climber or vine that is well represented in most of Europe, where it spread again after the glaciations. The "loro" (*Prunus lusitanica*) is the only tree that survives as a relict in some Iberian riversides, especially in the western part of the peninsula, particularly the Extremadura, and to a small extent in the Northeast. In other cases, the presence of Mediterranean laurel (*Laurus nobilis*) provides an indication of the previous existence of laurel forest. This species survives natively in Morocco, Italy, Portugal, Greece, the Mediterranean islands, and some areas of Spain, including the Parque Natural Los Alcornocales in the province of Cádiz and in coastal mountains, especially in the Girona Province of Catalonia, which keeps the best "lauredales", and isolated in the Valencia area. Cortegada Island in Galicia is famous for its vast forest of laurels, but this forest is not indigenous to the island, but originated in plantings after the native vegetation had been destroyed. The myrtle spread through North Africa. Tree Heath *(Erica arborea)* grows in southern Iberia, but without reaching the dimensions observed in the temperate evergreen forest or North Africa. The subspecies *Rhododendron ponticum baeticum* and/or *Rhamnus frangula baetica* still persist in humid microclimates, such as stream valleys, in Spain's Baetic Cordillera, in the Portuguese mountains of Monchique, and the Rif Mountains of Morocco.

Although the Atlantic laurisilva is more abundant in the Macaronesian archipelagos, where the weather has fluctuated little since theTertiary, there are small representations and some species contribution to the oceanic and Mediterranean ecoregions of Europe,Asia minor and west and north of Africa, where microclimates in the coastal mountain ranges form inland "islands" favorable to the persistence of laurel forests. In some cases these were genuine islands in the Tertiary, and in some cases simply areas that remained ice-free. When the Strait of Gibraltar reclosed, the species repopulated toward the Iberian Peninsula to the north and were distributed along with other African species, but the seasonally drier and colder climate, prevented them reaching their previous extent. In Atlantic Europe, subtropical vegetation is interspersed with taxa from Europe and North African in bioclimatic enclaves such as Monchique, Sintra, and the coastal mountains from Cadiz to Algeciras. In the Mediterranean region, remnant laurel forest is present in some islands of the Aegean Sea, on the Black Sea coast of Iran and Turkey, including the laurifolia castanopsis and true laurus forests, associated with *Prunus laurocerasus*, and conifers such as *Taxus baccata*, *Cedrus atlantica*, and *Abies pinsapo*.

In Europe the laurel forest has been badly damaged by timber harvesting, by fire (both accidental and deliberate to open fields for crops), by the introduction of exotic animal and plant species that have displaced the original cover, and by replacement with arable fields, exotic timber plantations, cattle pastures, and golf courses and tourist facilities. Most of the biota is in serious danger of extinction. The laurel forest flora are usually strong and vigorous, so the forest regenerates easily; its decline is due to external forces.

NEPAL

In the Himalayas, in Nepal, subtropical forest consists of species such as *Schima wallichii*, *Castanopsis indica*, and *Castanopsis tribuloides* in relatively humid areas. Some common forest types in this region include *Castanopsis tribuloides* mixed with *Schima wallichi, Rhododendron* spp., *Lyonia ovalifolia, Eurya acuminata*, and *Quercus glauca*;*Castanopsis*-Laurales forest with *Symplocas* spp.; *Alnus nepalensis* forests; *Schima wallichii-Castanopsis indica* hygrophile forest; *Schima-Pinus* forest; *Pinus roxburghii* forests with *Phyllanthus emblica. Semicarpus anacardium*, *Rhododendron arboreum* and *Lyoma ovalifolia; Schima-Lagestromea parviflora* forest, *Quercus lamellosa* forest with*Quercus lenata* and 'Quercus glauca; *Castanopsis* forests with *Castanopsis hystrix* and Lauraceae.

SOUTHERN INDIA

Laurel forests are also prevalent in the montane rain forests of the South Western Ghats in southern India.

SRI LANKA

Laurel forest occurs in the montane rain forest of Sri Lanka.

AFRICA

The Afromontane laurel forests describe the plant and animal species common to the mountains of Africa and the southern Arabian Peninsula. The afromontane regions of Africa are discontinuous, separated from each other by lowlands, resembling a series of islands in distribution. Patches of forest with Afromontane floristic affinities occur all along the mountain chains. Afromontane communities occur above 1500–2000 meters elevation near the equator, and as low as 300 meters elevation in the Knysna-Amatole montane forests of South Africa. Afromontane forests are cool and humid. Rainfall is generally greater than 700 mm per year, and can exceed 2000 mm in some regions, occurring throughout the year or during winter or summer, depending on the region. Temperatures can be extreme at some of the higher altitudes, where snowfalls may occasionally occur.

In Subsaharan Africa, laurel forests are found in the Cameroon Highlands forests along the border of Nigeria and Cameroon, along the East African Highlands, a long chain of mountains extending from the Ethiopian Highlands around the African Great Lakes to South Africa, in the Highlands of Madagascar, and in the montane zone of the São Tomé, Príncipe, and Annobón moist lowland forests. These scattered highland laurophyll forests of Africa are similar to one another in species composition (known as the Afromontaneflora), and distinct from the flora of the surrounding lowlands.

The main species of the Afromontane forests include the broadleaf canopy trees in *Beilschmiedia* genus, with *Apodytes dimidiata*, *Ilex mitis*, *Nuxia congesta*, *N. floribunda*,*Kiggelaria africana*, *Prunus africana*, *Rapanea melanophloeos*, *Halleria lucida*, *Ocotea bullata*, and*Xymalos monospora*, along with the emergent conifers *Podocarpus latifolius*and *Afrocarpus falcatus*. Species composition of the Subsaharan laurel forests differs from that of Eurasia. Trees of the Laurel family are less prominent, limited to *Ocotea* or*Beilschmiedia* due to exceptional biological and paleoecological interest and the enormous biodiversity mostly but with many endemic species, and the members of the beech family (Fagaceae) are absent.

Trees can be up to 30 or 40 meters (98 or 131 ft) tall and distinct strata of emergent trees, canopy trees, and shrub and herb layers are present. Tree species include: Real Yellowwood (*Podocarpus latifolius*), Outeniqua Yellowwood (*Podocarpus falcatus*), White Witchhazel (*Trichocladus ellipticus*), *Rhus chirendensis*, *Curtisia dentata*, *Calodendrum capense*,*Apodytes dimidiata*,*Halleria lucida*, *llex mitis*, *Kiggelaria africana*, *Nuxia floribunda*, *Xymalos monospora*, and *Ocotea bullata*. Shrubs and climbers are common and include: Common Spikethorn (*Maytenus heterophylla*), Cat-thorn (*Scutia myrtina*), Numnum (*Carissa bispinosa*), *Secamone alpinii*, *Canthium ciliatum*, *Rhoicissus tridentata*,*Zanthoxylum capense*, and *Burchellia bubalina*. In the undergrowth grasses, herbs and ferns may be locally common: Basketgrass (*Oplismenus hirtellus*), Bushman Grass (*Stipa dregeana* var. *elongata*), Pigs-ears (*Centella asiatica*), *Cyperus albostriatus*, *Polypodium polypodioides*, *Polystichum tuctuosum*, *Streptocarpus rexii*, and *Plectranthus* spp. Ferns, shrubs and small trees such as Cape Beech (*Rapanea melanophloeos*) are often abundant along the forest edges.

USA SOUTHEAST STATES

Laurel forests occur in the far southern portions of the southeastern United States from South Carolina to Florida and west to Texas. In most cases, these Laurel forests are concentrated on the drier coastal plains and coastal margins along the Gulf and far south Atlantic states. Despite being located in a humid climate zone, much of the broadleaf Laurel forests in the Southeast USA are semi-sclerophyll in character. The semi-sclerophyll character is due (in part) to the sandy soils and often periodic semi-arid nature of the climate. As one moves south (Florida), the sclerophyll character slowly declines and a more tropical and wet-dry character develops, and much of the laurel forest gives way to (tropical savanna).

In some areas evergreen forests are dominated by species of Live Oak (Quercus virginiana), Laurel Oak (Quercus hemisphaerica), Southern Magnolia (Magnolia grandiflora), Red Bay Persea borbonia, Cabbage Palm (Sabal palmetto), and Sweetbay Magnolia (Magnolia virginiana). Many non-native species have become naturalized and freely mix with native broadleaved evergreens in upland areas just above the coastal plain.

The lower shrub layer of the evergreen forests is often mixed with other evergreen species from the palm family (Rhapidophyllum hystrix), Bush palmetto(Sabal minor),and Saw Palmetto (Serenoa repens), and several species in the Ilex family, including (Ilex glabra), (Dahoon Holly), and (Yaupon Holly), as well as Wax Myrtle (Myrica) are present. Several species of Yucca are native as well to the drier sandy coastal scrub environment of the region, including (Yucca aloifolia), (Yucca filamentosa), and (Yucca gloriosa). These subtropical forests lie between the temperate deciduous and conifer forests to the north and the true tropical savanna and tropical mangrove and hardwood hammocks of deep south Florida.

USA ANCIENT CALIFORNIA

During the Miocene, oak-laurel forests were found in Central and Southern California. Typical tree species included oaks ancestral to present-day California oaks, as well as an assemblage of trees from the Laurel family, including *Nectandra*, *Ocotea*, *Persea*, and *Umbellularia*. Only one native species from the Laurel family (Lauraceae),*Umbellularia californica,* remains in California today.

CENTRAL AMERICA

The laurel forest is the most common Central American temperate evergreen cloud forest type. They are found in mountainous areas of southern Mexico and almost all Central American countries, normally more than 1,000 m above sea level. Tree species include evergreen oaks, members of the Laurel family, and species of *Weinmannia*, *Drimys*, and*Magnolia*. The cloud forest of Sierra de las Minas, Guatemala, is the largest in Central America. In some areas of southeastern Honduras there are cloud forests, the largest located near the border with Nicaragua.

In Nicaragua the cloud forests are found in the border zone with Honduras, and most were cleared to grow coffee. There are still some temperate evergreen hills in the north. The only cloud forest in the Pacific coastal zone of Central America is on the Mombacho volcano in Nicaragua. In Costa Rica there are laurisilvas in the "Cordillera de Tilarán" and Volcán Arenal, called Monteverde, also in the Cordillera de Talamanca.

LAUREL FOREST ECOREGIONS IN MEXICO AND CENTRAL AMERICA

- Central American montane forests
- Chiapas montane forests
- Chimalpas montane forests
- Oaxacan montane forests
- Talamancan montane forests
- Veracruz montane forests

TROPICAL ANDES

The Yungas are typically evergreen forests or jungles, and multi-species, which often contain many species of the laurel forest. They occur discontinuously from Venezuela to northwestern Argentina including in Brazil, Bolivia, Chile, Colombia, Ecuador, and Peru, usually in the Sub-Andean Sierras. The forest relief is varied and in places where the Andes meet the Amazon, it includes steeply sloped areas. Characteristic of this region are deep ravines formed by the rivers, such as that of the Tarma River descending to the San Ramon Valley, or the Urubamba River as it passes through Machu Picchu. Many of the Yungas are degraded or are forests in recovery that have not yet reached theirclimax vegetation.

SOUTHEASTERN SOUTH AMERICA

The laurel forests of the region are known as the *Laurisilva Misionera*, after Argentina's Misiones Province. The Araucaria moist forests occupy a portion of the highlands of southern Brazil, extending into northeastern Argentina. The forest canopy includes species of Lauraceae (*Ocotea pretiosa* and *O. catharinense*), Myrtaceae (*Campomanesia xanthocarpa*), and Leguminosae (*Parapiptadenia rigida*), with an emergent layer of the conifer Brazilian Araucaria (*Araucaria angustifolia*) reaching up to 45 meters (148 ft) in height. The subtropical Serra do Mar coastal forests along the southern coast of Brazil have a tree canopy of Lauraceae and Myrtaceae, with emergent trees of Leguminaceae, and a rich diversity of bromeliads and trees and shrubs of family Melastomaceae. The inland Alto Paraná Atlantic forests, which occupy portions of the Brazilian Highlands in southern Brazil and adjacent parts of Argentina and Paraguay, are semi-deciduous.

CENTRAL CHILE

The Valdivian temperate rain forests, or *Laurisilva Valdiviana*, occupy southern Chile and Argentina from the Pacific Ocean to the Andes between 38° and 45° latitude. Rainfall is abundant, from 1500 to 5000 mm according to locality, distributed throughout the year, but with some subhumid Mediterranean climate influence for 3–4 months in summer. The temperatures are sufficiently invariant and mild, with no month falling below 5 °C, and the warmest month below 22 °C.

AUSTRALIA, NEW CALEDONIA AND NEW ZEALAND

Laurel forest appears on mountains of the coastal strip of New South Wales in Australia, New Guinea, New Caledonia,Tasmania, and New Zealand. The laurel forests of Australia, Tasmania, and New Zealand are home to species related to those in the Valdivian laurel forests, including Southern Beech (*Nothofagus*, fossils of which have recently been found in Antarctica) through the connection of the Antarctic flora. Other typical flora include Winteraceae, Myrtaceae, Southern Sassafras (Atherospermataceae), conifers of Araucariaceae, Podocarpaceae, and Cupressaceae, and tree ferns.

New Caledonia was an ancient fragment of the supercontinent Gondwana. Unlike many of the Pacific Islands, which are of relatively recent volcanic origin, New Caledonia is part of Zealandia, a fragment of the ancient Gondwana that separated from Australia 60–85 million years ago, and the ridge linking New Caledonia to New Zealand has been deeply submerged for millions of years. This isolated New Caledonia from the rest of the world's landmasses, preserving a snapshot of Gondwanan forests. New Caledonia and New Zealand are separated by continental drift of Australia 85 million years ago. The islands still shelter an extraordinary diversity of endemic plants and animals of Gondwanan origin that have later spread to the southern continents.

The laurel forest of Australia, New Caledonia (*Adenodaphne*), and New Zealand have a number of species related to those of the Valdivian laurel forest, through the connection of the Antarctic flora of gymnosperms like the podocarpus and deciduous Nothofagus. *Beilschmiedia tawa* is often the dominant canopy species of genus *Beilschmiedia* in lowland laurel forests in the North Island and the northeast of the South Island, but will also often form the subcanopy in primary forests throughout the country in these areas, with podocarps such as Kahikatea, Matai, Miro and Rimu.

Genus *Beilschmiedia* are trees and shrubs widespread in tropical Asia, Africa, Australia, New Zealand, Central America, the Caribbean, and South America as far south as Chile. In the Corynocarpus family, *Corynocarpus laevigatus* is called laurel of New Zealand, while *Laurelia novae-zelandiae* belongs to the same genus as *Laurelia sempervirens*. The tree niaouli grows in Australia, New Caledonia, and Papua.

New Caledonia lies at the northern end of the ancient continent Zealandia, while New Zealand rises at the plate boundary that bisects it. These land masses are two outposts of the Antarctic flora, including Araucarias and Podocarps. At Curio Bay, fossilized logs can be seen of trees closely related to modern Kauri and Norfolk Pine that grew on Zealandia about 180 million years ago during the Jurassic period, before it split from Gondwana.

During glacial periods more of Zealandia became a terrestrial rather than a marine environment. Zealandia was originally thought to have no native land mammals, but a recent discovery in 2006 of a fossil mammal jaw from the Miocene in the Otago region shows otherwise.

The New Guinea and Northern Australian ecoregions are closely related. Over time Australia drifted north and became drier; the humid Antarctic flora from Gondwana retreated to the east coast and Tasmania, while the rest of Australia became dominated by sclerophyll forest and xeric shrubs and grasses. Humans arrived in Australia 50–60,000 years ago, and used fire to reshape the vegetation of the continent; as a result, the Antarctic flora, also known as the *Rainforest flora* in Australia, retreated to a few isolated areas composing less than 2% of Australia's land area.

NEW GUINEA

The eastern end of Malesia, including New Guinea and the Aru Islands of eastern Indonesia, is linked to Australia by a shallow continental shelf, and shares many marsupialmammal and bird taxa with Australia. New Guinea also has many additional elements of the Antarctic flora, including southern beech (*Nothofagus*) and Eucalypts. New Guinea has the highest mountains in Malesia, and vegetation ranges from tropical lowland forest to tundra.

The highlands of New Guinea and New Britain are home to montane laurel forests, from about 1000 up to 2500 meters elevation. These forests include species typical of both Northern Hemisphere laurel forests, including *Lithocarpus, Ilex,* and Lauraceae, and Southern Hemisphere laurel forests, including Southern Beech *Nothofagus*, *Araucaria*,Podocarps, and trees of the Myrtle family (Myrtaceae). New Guinea and Northern Australia are closely related. Around 40 million years ago, the Indo-Australian tectonic plate began to split apart from the ancient supercontinent Gondwana. As it collided with the Pacific Plate on its northward journey, the high mountain ranges of central New Guinea emerged around 5 million years ago. In the lee of this collision zone, the ancient rock formations of what is now Cape York Peninsula remained largely undisturbed.

TEMPERATE CONIFEROUS FOREST

Temperate coniferous forest is a terrestrial biome found in temperate regions of the world with warm summers and cool winters and adequate rainfall to sustain a forest. In most temperate coniferous forests, evergreen conifers predominate, while some are a mix of conifers and broadleaf evergreen trees and/or broadleaf deciduous trees. Temperate evergreen forests are common in the coastal areas of regions that have mild winters and heavy rainfall, or inland in drier climates or mountain areas. Coniferous forests can be found in the United States, Canada, Europe, and Asia. Many species of tree inhabit these forests including cedar, cypress, Douglas fir, fir, juniper,kauri, pine, podocarpus, spruce, redwood and yew. The understory also contains a wide variety of herbaceous and shrub species.

Fig. A pine forest is an example of a temperate coniferous forest

Structurally, these forests are rather simple, generally consisting of two layers: an overstory and understory. Some forests may support an intermediate layer of shrubs. Pine forests support a herbaceous understory that is generally dominated by grasses and herbaceous perennials, and are often subject to ecologically important wildfires.

Temperate rain forests occur only in seven regions around the world: the Pacific temperate rain forests of the Pacific Northwest, theValdivian temperate rain forests of southwestern South America, the rain forests of New Zealand and Tasmania, northwest Europe (small pockets in Ireland, Scotland, Wales, Iceland and a somewhat larger area in Norway), southern Japan, and the eastern Black Sea-Caspian Sea region of Turkey and Georgia to northern Iran. The moist conditions of temperate rain forests generally support an understory of mosses, ferns and some shrubs. Temperate rain forests can be temperate coniferous forests or temperate broadleaf and mixed forests.

SPECIES

The temperate coniferous rain forests sustain the highest levels of biomass in any terrestrial ecosystem and are notable for trees of massive proportions, including coast redwood (*Sequoia sempervirens*), Douglas fir (*Pseudotsuga menziesii*), Sitka spruce (*Picea sitchensis*), Alerce (*Fitzroya cupressoides*) and kauri (*Agathis australis*). These forests are quite rare, occurring in small areas of North America, southwestern South America and northern New Zealand. The Klamath-Siskiyou forests of northwesternCalifornia and southwestern Oregon is known for its rich variety of plant and animal species, including many endemic species.

The wildlife here includes rich fur bearing animals such as mink, silver fox, lynx, sable, and beaver. The leaves on a coniferous tree are needle like. The leaves have a waxy coating to help the leaves from being damaged.

THE BENEFITS OF EVERGREEN CONIFERS FOR ENERGY CONSERVATION

More than 100 species of native and introduced evergreen conifers are found in the forests and cities of North America. These conifer species have evergreen needles and bear seed cones; however, they vary considerably in where they are found, how big they grow, how fast they grow, and how suitable they are for home landscapes. Evergreen conifers provide both winter and summer energy conservation benefits. If appropriately planted and cared for, they can reduce chilling winds in the winter and provide shade in the summer.

Because evergreen conifers keep their foliage year-round, they are well-suited for use in windbreaks to abate harsh winter winds. A windbreak is a dense grouping of trees, often situated in several parallel rows, that is planted adjacent to a house, building, or outdoor space to intercept and slow prevailing winds. By slowing wind speed, windbreaks help conserve energy used to heat buildings. If situated correctly, these same trees can also provide shade in the summer.

Much like the human body, a building is kept warm by an energy source that elevates the core temperature, then circulates the warmth around the extremities. The drywall, fiberglass, brick, and vinyl that cover the walls are like a coat, helping to insulate the building from heat loss. When cold air passes across the exterior walls, heat is transferred from the wall surfaces to the cold air. This is analogous to windchill for people. A windbreak of trees slows down the wind speed so that a stable layer of slightly warmer air forms around the building exterior, minimizing the temperature difference between the building and the outside air. Less heat is is then "pulled" from the building to warm the surrounding air and there is less demand on the building's heating system.

To maximize the heat conservation benefit of a conifer windbreak, the proper species must be selected and placed in the proper orientation and distance from the building. For best effect, the trees chosen for a windbreak must attain a height similar to the building.

Most conifers are capable of growing to that stature. In fact, many conifers may grow too large for typical residential landscapes, which can lead to conflicts with driveways, sidewalks, rooftops, and power lines. Moreover, conifers may differ in their requirements for soil fertility and drainage. Most pines are adapted to dry, infertile soil, but some species only thrive on cool, rich soils. Measure the space in which your trees can grow and have a soil sample tested prior to selecting the conifer species for your windbreak. For best effect, a conifer windbreak must be situated to block the prevailing winter wind. For much of North America, these winds prevail from the northwest, although local differences may exist (Diagram 1). Information on prevailing wind patterns in your vicinity may be obtained from the National Weather Service.

Windbreaks are most effective when the trees are planted relatively close to the building being sheltered. A good rule of thumb is a distance no more than the mature height of the trees being planted. At greater distances, the windbreak becomes less effective. However, do not plant the windbreak too close to the building, because this can lead to nuisances from errant branches and roots. Evergreen conifers can also help conserve energy in summertime by casting shade upon houses, garages, and other buildings. By casting shade, the trees are intercepting solar radiation that would otherwise strike the building surfaces and warm them. By keeping building surfaces cool, less heat is transferred to the inside air. Therefore, the air conditioning system has to be run less frequently to keep the building at a comfortable indoor temperature.

To maximize the shading benefit, trees should be planted on the west and south aspect of the building to intercept the hot afternoon summer sun. In northerly latitudes, it is undesirable to plant large evergreen conifers on the south aspect of a building. The dense shade that they cast in winter keeps the sun from heating the building. Without this free natural heat source, more energy must be consumed to warm the building. Deciduous shade trees are suitable for the south aspect because they shed their leaves in winter.

In summary, homeowners should balance their aesthetic preferences, energy conservation needs, and anticipated tree growth when choosing evergreen conifers for their home landscapes. Diligent research regarding the best location to plant on site should also be undertaken. For maximum energy conservation benefits, conifers should be planted to block prevailing winter winds from the north and west, yet kept away from the south aspect to take full advantage of passive solar heating in winter.

2

Tropical Dry Evergreen Forests of India

In the tropics, changes in the quantity and distribution of rainfall along with temperature and the length of the dry season gradually alter the vegetation formation . The pronounced seasonality in rainfall distribution with several months of drought result in seasonally dry forests in tropical regions . As dry forests have a broad climatic range, transitional forest ecosystems like grasslands, savannas, scrub, and thorn woodlands are often considered during the vegetation assessments . Many times, these dry forests in the varying climatic regimes differ in their forest structure and physiognomy. The prevailing tropical dissymmetric climate regime on the Coromandel coast of southern peninsular India supports a unique type of vegetation named tropical dry evergreen forest (TDEF) . The Coromandel coastal plains extend about 80-100 km inland .

The tropical dry evergreen forest (TDEF) type, first described as a low forest of 9 to 12 m high, forms, however, a complete canopy comprising small, coriaceous-leaved evergreen trees of short boles and spreading crowns with some deciduous emergents, without marked differentiation of canopy layers . Floristically, it is distinguished by a fair representation of characteristic and preferential species, exclusively or mostly confined to this vegetation type. The tropical dry evergreen forests on the Coromandel coast of India, which occur as patches, are short-statured, largely three-layered, tree-dominated evergreen forests with a sparse and patchy ground flora .

Each region of the world has a vegetation type that has, over countless eons, evolved as the plant community most suited to the environmental conditions of the area. The Tropical Dry Evergreen Forest (TDEF) is the indigenous forest of the coastal seaboard of South East India. Historically the forest extended from Vishakapatanam to Ramanathapuram as a belt of vegetation between 30 and 50 km wide, bordered on one side by the sea and on the other side by a forest that becomes increasingly deciduous as one moves inland.

It contains over 160 woody species of which around 70 are found within the pristine climax forest. This is predominantly composed of trees and shrubs that have thick dark green foliage throughout the year. There are six vegetative elements: trees, shrubs, lianas, epiphytes, herbs, and tuberous species. In the pristine state these components weave together to form a complex diverse habitat that is home to a myriad of animal species, mammals, birds, reptiles, amphibians, insects, as well as a host of microbes.

When one includes all of the herbaceous species that grow in a variety of ecological niches within the range of the forest the number of species approaches 1000, of which over 600 have a recorded use for mankind either medicinally, culturally or in religious rituals.

The relevance of the forest today lies both within its vast botanical wealth, and also its ability to ameliorate the environmental conditions that are steadily deteriorating due to the expanding population and increase of consumer lifestyles. However there is hardly any of this forest that remains free from human interference, the vast majority of forests in the area are little more than degraded thorny thickets, lacking the inherent nobility of the climax vegetation. It is the intention of this booklet to draw attention to the TDEF and put forward the case for its conservation.

THE VEGETATION

The forest when in its pristine state is a tightly woven matrix of vegetation about 6m high with the occasional emergent rising above the canopy to 10 meters. The trees and shrubs of the forest exhibit many characteristics that are similar, indicating a form of convergent evolution. They are evergreen, responding to the rains with a flush of new leaves. The leaves are mostly simple, thick and waxy, with a size around 6cm by 3cm. The flowers are small, 1 cm in diameter, often white with a perfume, the season for them is generally between February and August. The fruits are small and fleshy, ripening mainly between April and September. The habit of the trees is generally to have around two meters of clean trunk, and then to branch. In some species buttressing is found. The overall height of the trees is between 4 and 7 meters. The trees are slow growing, the wood is dense and hard, in the main thorns are absent although in this particular instance there are some notable exceptions.

Some of the common species exhibiting all or most of these characteristics are Atalantia monophylla, Diospyros ebenum, Drypetes sepiaria, Garcinia spicata, Glycosmis mauritiana, Ixora pavetta, Lepisanthes tetraphylla, Manilkara hexandra, Memecylon umbellatum, Syzygium cumini.

Pterospermum suberifolium is also an integral part of the climax forest, but it is the common emergent, and has many different physiological features which are probably due to the environmental conditions it experiences above the canopy. Its leaves are larger, thinner, white hairs beneath, the flowers are larger and the seeds are wind blown. Other important vegetative components of the forest are the climbers/liana (eg. Combretum ovalifolium, Capparis zeylanica) and also the orchids that are both epiphytic (eg. Vandia tesselata) and pseudobulbous (eg. Eulophia epidendraea). The lianas can be up to 30cm in girth and can extend over 10 or more tree canopies. They do not follow the vegetative characteristics of the trees and shrubs, and this might indicate that they are signs of disturbance. The orchids are rarely found in the forests now as these are the most vulnerable species to disturbance.

THE ANIMALS

In present times the highly degraded state of the forest, and the high human population pressure means that the larger predatory mammals can no longer be found in the region. However it is still surprising the number of small mammals that hang on especially in the larger reserve forests of the area. It is thought that the pangolin, the honey badger are still to be found, and in much larger numbers the porcupine, the hedgehog, the fox and the mongooses reside in the forests. In specific areas (Point Calimere) the black buck and the chital are found, and the bonnet macaque is resident in most forests.

The reptiles of the forest include the monitor lizard, the chameleon, and other lizards, as well as 19 species of snake, including the 4 venomous species of the plains, the cobra, the krait, and the two vipers.

The bird population of the forests has been estimated around 80 species, and in certain areas the trees provide valuable roosting sites for water birds at night. The bird population is made up of residents, who stay through out the year, and migrants who arrive with the rains and who often take advantage of the burgeoning insect population that is concurrent with the monsoon.

Less glamorous, but perhaps more important to the healthy functioning of an ecosystem is the insect population, and the other microfauna found within the leaf litter and humus layer. Little is known about these animals, but simple observations indicate a wealth that is yet to be studied. For example in the leaf litter of an undisturbed forest in a sacred grove, from an area of one square meter 5 species of spiders, 2 species of roaches, 2 species of ants, a bug and a cricket were found, all accounting for 368 individual specimens!

THE FOREST AND PEOPLE

The forests of the region have been intimately connected to humans for countless centuries. Prior to the Roman times, when trade was exchanged with the coastal communities, civilization has been recorded in the area. And even before these times the coastal area would have attracted fishermen and the light sandy soils would have been accessible to primitive agricultural tools and crops.

In recent history, that is the last two hundred years these forests have been treated as a harvestable resource, and this has lead to their degradation.

However even in today's highly degraded state they still remain a vast repository of natural resources. It has been recorded that almost all the local species have a use. Four hundred species of trees, shrubs and herbs are utilized in traditional medicine practiced by the villagers from householders and midwives to specialists treating bone fractures, poisonous bites and eye ailments.

In India, modern medical health systems can only offer primary health care for up to 30% of the rural population. Plant based indigenous health systems can help with providing real health security as they are very much alive and are evolving and adapting to modern needs. This information on bio-resource use represents a valuable record for posterity and also a phenomenal bank of information that can be drawn upon to resolve current predicaments.

As well as providing possibilities for the material and medical needs of the population the forest also provides for the spiritual. In fact the only pristine areas of forest left are around temples as sacred groves. The temples are often dedicated to the god Iyanar, and although all of these groves are suffering from disturbance, within from pilgrims and at the edges from surrounding farmers they still provide a setting for many stories that are told concerning the past.

THE FUTURE OF THE FOREST

The possibility to conserve, protect and actually restore the Tropical Dry Evergreen Forest still exists today. Work has been carried out, researching the forest, establishing which species go to make up the matrix of the mature forest, collecting information as to when each species flowers and fruits, and developing techniques to germinate and raise the species in nurseries to plant out either as enrichment plantings in already existing forest, or to rehabilitate bare and barren land.

The next step needs to be taken, but still many questions need to be addressed. The future of this forest is dependant upon the local population rediscovering the respect that was once held for the forest, recognizing the role that it can play in the needs and necessities of today's ever changing society. We need to work within these areas, ensure that the basic needs of the people are taken care for, and help to strengthen the community so that they can take care of the forest themselves.

At a principle, fundamental level the forest provides for the foundation of the surrounding communities. It can enhance the water holding capacities of the soil, ensuring that the ground water resource is secure. It provides a home for the pollinators and predators of pests that can help support sustainable agricultural practices. And perhaps most importantly it provides a link with where we came from, the way nature developed a covering for this particular land, and where our ancestors walked and evolved their ideas about who we are and the way we should conduct ourselves.

Strategies need to be developed on many levels, to address the different possibilities for the conservation of the TDEF. Ways of involving the whole community from government, through business, and on to academic institutions and NGO's, and all the way via schools to the local public. Be it management and development plans for the existing reserve forests, encouraging corporations to plant TDEF within their compounds, designing forests for planting with the local communities and schools, or simply creating an awareness about the existence of a beautiful forest under threat within the population at large. The work needs to commence soon, so that we can get there sooner rather than later.

DISTRIBUTION OF TROPICAL DRY EVERGREEN FOREST

Dry evergreen forests have also been reported elsewhere in the tropics, along with aspects researched therein. The information is a result of a systematic review that has been made possible through extensive survey of literatures that report the occurrence of dry evergreen forests, either as a vegetation formation or as a forest type. There are no unified features for this rare and unique forest type and it has been chosen based on local climatic, biotic and edaphic factors, which influence the forest's physiognomy, stand structure, species composition, and dynamics.

Table. Distribution of tropical dry evergreen forests in the tropics and aspects studied therein. (* Cl-Climate; Sl-soil; Veg-vegetation structure; FC-floristic composition; Dyn-dynamics; Phy-physiology; N.Cy-nutrient cycling; Repr-reproductive ecology; FU-forest utilization; T&D-threats and disturbance; Con-conservation)

Location	Aspects studied*											Reference
	Cl	Sl	Veg	FC	Dyn	Phy	N.Cy	Repr	FU	T&D	Con	
Tropical America												
Antigua	v	v	v	v								16
Bahamas	v	v	v	v						v		10-13
British Guiana	v	v	v	v								15-16
Jamaica	v	v	v	v								17-19
Trinidad	v	v	v	v								20
Tobago	v	v	v	v								21
Africa												
Ethiopian highlands	v	v	v	v					v	v	v	22
Tanzania	v	v	v	v			v				v	23-24
Zambia	v	v	v	v						v		25-26
Asia												
Thailand	v	v	v	v	V	v	v		v	v	v	27-31
Sri Lanka	v	v	v	v	V	v	v		v	v	v	32-34
India	v	v	v	v	V		v	v	v	v	v	5, 35-46, 49-52, 54-58, 64

Beard (1955) recognized six dry evergreen formations in tropical America, which are formed due to strong winds and/or excessively freely draining soil, whereas the rainfall regime there is not of dissymmetric type. In tropical America, TDEF occurs in the North Andros islands of the Bahamas as a "coppice community," which is a dense, close-canopied, broad-leaved evergreen forest, and in British Guiana, Antigua, Jamaica, Trinidad, and Tobago. In Africa, TDEF is reported as montane evergreen scrubland vegetation in multi-storied form in the highlands of Ethiopia, and in Tanzania and north-eastern Zambia as scattered patches of closed canopy of evergreen shrubs of 15-25 m tall (locally, known as "Matechi").

Dry evergreen forests as a closed-canopy evergreen forest type, with 25-30 m of mean canopy height, are widespread in the regions of Thailand that receive not more than 1,200 mm mean annual rainfall, with 4-6 dry months. In Sri Lanka, dry evergreen forest is typical and a dominant vegetation type in the dry zone regions in the northern and eastern plains, which cover 80 percent of the island area.

InIndia, this vegetation is confined to the Coromandel (east) coast region. Some patches of dry evergreen forests have also been recorded in the Sirumalai hills, Kolli hills, Shervarayan hills and Chitteri hills of southern Eastern Ghats. However, the climate and characteristic species of hill dry evergreen type are not the same as that of the coastal region.

In reality, most of the Indian TDEFs, with the exception of two large areas, namely the Kurumbaram section of the Marakanam Reserve Forest and the Point Calimere Wildlife Sanctuary, occur as patches of forest dotted along the Coromandel coast, and invariably protected as "sacred groves" based on the religious belief of the local people. This unique dry evergreen forest is relatively under-studied on aspects of structural and functional ecology, as compared to the tropical wet evergreen forests. The aim of this paper is to provide a consolidated account on plant biodiversity, structure and functional ecology, and bioresource potential, particularly of medicinal plants, and to emphasize the conservation need and significance of TDEFs on the Coromandel coast of peninsular India.

METHODS

Study area

Investigations on plant biodiversity and bioresource potential of 75 TDEF sites, which are concentrated in the Pondicherry (11°56' N and 79°53' E), Villupuram (11°93' N and 79°48' E), Cuddalore (11°43' N and 79°49' E) and Pudukottai (10°23' N and 78°52' E) districts on the Coromandel coast of peninsular India, were carried out. The areal extent of TDEF sites studied ranged from 0.5 ha to ~10 ha. The climate is tropical dissymmetric type with most rainfall received during the northeast monsoon (October-December) and very little and inconsistent rainfall in the southwest monsoon (June to September). The mean annual rainfall is 1,282, 1,079 and 1,033 mm in the nearest towns, namely Pondicherry, Cuddalore, and Pudukottai, respectively. The dry season lasts for six months (January to June), and receives less than 60 mm rainfall on monthly average. Mean annual maximum and minimum temperatures are 32.58°C and 24.51°C in Pondicherry, 22.75°C and 33.64°C in Cuddalore, and 33.4°C and 25.4°C in Pudukottai.

Data collection

Field data collection on species check listing and assessment of bioresource values in the 75 TDEF sites was conducted in about 125 man days during July 2006-January 2008. The dataset on woody plant diversity (for trees equal or greater than 10 cm girth at breast height, (1.3 m height from ground level), and all lianas =1 cm diameter measured at 1.3 m from the base of the stem), dynamics, and functional ecology reviewed here is based on the systematic investigations carried out in a total of 12 one-ha permanent plots over a decade on the Coromandel coast of peninsular India. All plant species were identified and confirmed to species level using regional floras. Voucher specimens were collected and deposited in the herbarium of the Department of Ecology, Pondicherry University.

Site disturbance scores were obtained by assessing various disturbances (on a 1-5 scale) which include site encroachment, distance from the human habitation, temple visitors' impact, cattle grazing, resource removal, width of approach road to temple, fragmentation, size of the temple, biological invasion, and frequency of peoples' visit to the temple. The summed disturbance score of each site was used for ranking the TDEF sites into three categories, viz relatively undisturbed, moderately disturbed and highly disturbed, for evaluating the conservation significance. The sites with low ranks experience least disturbance, while high ranks reveal a high level of anthropogenic disturbance in the site.

Medicinal plant resource use and traditional knowledge related to plant species of TDEFs were collected through a field-tested improved questionnaire and personal interviews with folk healers in their vernacular language. There were 47 informants (40 males and 7 females), who are folk healers by profession and part of the local folk healers association.

RESULTS AND DISCUSSION

Biodiversity

In a total of 75 TDEF sites on the Coromandel coast of peninsular India, 149 woody species that belonged to 122 genera and 49 families were enumerated. In addition to these, three important native herbaceous species occur there, which include the widely-distributed, colony-forming *Sansevieriaroxburghiana*, fairly distributed *Ecbolium viride*, and the rare *Amorphophallus sylvaticus*. Dominant families in Indian TDEFs include Euphorbiaceae, Rubiaceae with 11 species each, followed by Capparaceae, Mimosaceae, Fabaceae and Moraceae with 8 species each, while Alangiaceae, Barringtoniaceae and Burseraceae are represented by single species. Across

75 sites studied, species richness of woody plants ranged from 10 (Azhiyanilai and Thakkiripatti) to 69 species (Puthupet), and sites with more than 50 species include Puthupet (69 species), Kuzhanthaikuppam, Shanmuganathapuram and Oorani. The Morisita-Horn index for similarity of species composition between 75 TDEF sites (1.0 indicates total similarity) varied from 0 to 0.8 and only 10% of pairs had = 0.5, indicating the greater heterogeneity in the composition of species.

Fig.: Forest and interior stand view of relatively undisturbed (a&b), moderately disturbed (c&d) TDEF sites, landscape of TDEF with goat herding (e) and part of site converted to *Acacia* monoculture(f). a. Site OR-View of TDEF vegetation; b. RP-*Chloroxylon-Pterospermum* dominated stand; c. IT- Forest and sacred grove in entrance; d. MN-Inner stand view; e. KT-Herding goats inside the forest; f. OME-Converted to *Acacia leucophloea*(Mimosaceae) monoculture.

Among life forms, trees were dominant (102 species) representing 68% of the total species, while lianas formed 32% (47 species). Tree species richness at individual site ranged from 9 in Mettupatti to 36 species in Shanmuganathapuram. Mean species richness of lianas at each site was 30%. Maximum number of lianas was recorded at Puthupet (33 species), whereas lianas were virtually absent at Thakkiripatti. Some unique TDEF species such as *Pterospermum xylocarpum*, *Millusa montana*, *Polyalthia suberosa*, and*Alangium salvifolium* among trees, and *Olax scandens*, *Capparis rotundifolia*, *Pachygone ovata*, and*Mearua oblongifolia* among lianas occurred only in a few sites.

Earlier quantitative ecological inventory of plant biodiversity in 12 1-ha TDEF permanent plots has resulted in 86 tree species with a range of 19 to 35 species. A ubiquitous tree,*Memecylon umbellatum*, was the most dominant species, accounting for 32% of tree density, followed by*Tricalysia sphaerocarpa* (10.5%) and *Pterospermum canescens* (9.7%) in the tropical dry evergreen forests. A total of 44 liana species was inventoried with a range of 21-29 species ha^{-1} in the eight 1-ha plots. Among the lianas, *Combretum albidum* (19.2%), *Strychnos minor* (14%), and *Reissantia indica* (6.5%) were predominant species.

Although the 75 sites studied belong to the same TDEF type and are grossly homogeneous, they differ in forest stature; sites occurring on sandy soil with alluvium deposits are comparatively tall-statured (mean ht ~12 m; eg. TM, OR, AP, etc.) and those on red ferralitic hard compact soil are short- to medium-statured (mean ht <8 m; eg. TK, KP, MK, etc.). There is a wide variation in species composition of tree and liana species across TDEF sites, and interestingly each site is dominated by a different set of tree and liana species, which can be designated as "series," adding to the uniqueness of the studied TDEF sites (e.g., *Manilkara hexandra* in SV and *Memecyclon umbellatum-Tricalysia sphaerocarpa-Diospyros ebenum* in KK among trees; *Strychnos minor-Jasminum angustifolium* in PP and *Reissantia indica-Strychnos minor-Combretum albidum* in OR among lianas).

Forest Structure, Growth, and Dynamics

Out of 149 species, 75 are evergreen (50%), followed by deciduous (45 species, 30%) and brevi-deciduous species [species with brief deciduous period followed by synchronous leaf-flushing, e.g., *Pterospermum canescens*] (29 species; 20%). Among the 26 most common species, which occurred inmore than 30 sites, 54% were evergreen, 31% deciduous, and 15% brevi-deciduous. Similar results were reported in the study conducted in 43 TDEF sacred grove sites that contained 48% to 85% evergreen species.

Among the three physiognomic groups, evergreenness was prominent among trees (49%) and lianas (53%). The naturally evolved assemblage of evergreen species in dry evergreen vegetation type may be related to leaching of nutrients from leaves and year-round leaf fall, which is characteristic of evergreen species that establish a more closed nutrient cycle in the forest. Litter production quantified in two TDEF sites, namely Kuzhanthaikuppam (KK) and Oorani (OR), revealed a year-round litterfall with unimodal summer peak. Leaf litter production amounted to 9.6 and 9 t ha^{-1} yr^{-1} at KK and OR, respectively, while the standing crop of total forest floor litter was 4.11 t ha^{-1} at KK and 4.86 t ha^{-1} at OR.

Plant population changes have been studied in seven TDEFs by measuring the tree growth, recruitment, and mortality rate over years. Forest growth determined as girth increment in TDEF sites ranged from 0.37 to 1.08 cm yr^{-1} for trees and 0.39 to 0.41 cm yr^{-1} for lianas over a three-year period (2003-2006). The tree recruitment rate ranged from 0.7 to 2.3% yr^{-1}, while the mortality rate ranged from 1 to 2.2% yr^{-1} in five TDEF sites studied over three years (2003-2006). More small trees (10-30 cm gbh) recruitment in the forest has been attributed to selective logging of trees of highest girth class (>150 cm gbh) for temple construction that allowed more canopy gaps and sunlight.

Above ground biomass of 10 TDEF sites was estimated between 39.69 and 170.02 Mg ha^{-1} with a mean of 102.15 Mg ha^{-1}.

Reproductive Ecology

Analysis of qualitative reproductive traits of TDEF species revealed that many species had rotate-type, white-colored, scented flowers with nectar and pollen as rewards. Drupe and berry were the common fruit types and were found in black and red color, respectively. A strong association between the qualitative reproductive traits and pollination and dispersal spectrum among the TDEF species has been demonstrated. Phenological observations on TDEF species revealed a seasonal and unimodal flowering pattern with dry season peak at the community level.

A similar pattern of dry season flowering peak is also reported in other tropical seasonal dry forests. Many species exhibited annual flowering except a few species such as *Garcinia spicata*, *Reissantia indica*, *Dodonaea angustifolia*, etc., which exhibited a sub-annual pattern. The deciduous species (e.g., *Lannea coromandelica*, *Buteamonosperma*) displayed flowering and leaf shedding in dry summer. Species that flower during the high temperature and less rainfall attract diverse insects, while bee pollination was the prevalent mode of pollination system (68% of species) in the TDEF.

Fig.: Some characteristic tree and liana species of TDEFs a. *Memecylon umbellatum* (Melastomataceae)-predominant tree of TDEFs; b. *Pterospermum canescens* (Sterculiaceae) with woody capsule-common, lofty treeendemic to Coromandel coast TDEFs; c. *Aglaia elaegnoidea* (Meliaceae) - vertebrate-dispersed berries; d. *Eugenia bracteata* (Myrtaceae)-flowering twig; e. *Hugonia mystax* (Linaceae)-hook climber; f. *Capparis brevispina*(Capparaceae)-common thorny scrambler.

A bimodal fruiting pattern with a major peak in the dry season and a minor one in the early wet season was exhibited at the community level. There was year-round fruit production without a clear seasonality, but fruiting patterns at species level showed pronounced seasonality, which is in conformity with other seasonal forests. The patterns of unimodal flowering and year-round fruiting pattern are common to seasonal dry tropical forests, and these patterns have evolved according to local climatic factors (temperature, rainfall, number of dry months) along with ecological factors like availability of pollinators and dispersers. Most trees in our TDEFs flower (63%) and fruit (50%) during the dry period, whereas lianas had major flowering (77%) and fruiting (57%) activity in the late wet to dry season of the year. Many species are dispersed by animals, and had fruiting peak during the late dry season, which enables seed germination and rapid seedling establishment at the onset of the rainy season. The community-level fruit production in TDEF sites averaged 757 $kg^{-1}ha^{-1}yr^{-1}$.

Our understanding of species biology, particularly reproductive ecology, is still in an infant stage, and future directions for promising research include: (a) the environmental cues, which influence the phenological pattern; (b) the reproductive biology of important species (trees: *Pterospermum xylocarpum*,*Casearia elliptica*, *Aglaia elaegnoidea*; lianas: *Tiliacora acuminata* and *Strychnos minor*; herbs:*Sansevieria roxburghiana* and *Amorphophallus sylvaticus*) and also of dioecious tree species to assess the minimum viable population; (c) the extent of specialization in plant-pollinator interactions; (d) the level of inbreeding within and among species; (e) the impact of habitat fragmentation on pollination and fruit dispersal; and (f) genetic diversity analysis of polymorphic species such as *Memecylon umbellatum*,*Pterospermum xylocarpum*, etc.

IMPLICATIONS FOR CONSERVATION

Bioresource value

A total of 150 plant species that belonged to 57 families are reported to have medicinal value. They include 41 trees, 18 lianas, 14 shrubs, 10 herbaceous climbers, and 66 herbs. *Andrographis paniculata*,*Phyllanthus amarus*, *Gymnema sylvestre*, *Solanum nigrum*, and *S. trilobatum* are commonly used. A few characteristic/important medicinal species. The proportion of plant species used for medicinal purpose classed by plant parts include leaves (41%), fruits and seeds (14%), bark (12%), root (8%), latex (7%), whole plant (6%), and flower and bulbs (1%). Traditional healers use these plants for curing more than 52 ailments, mainly poisonous bites (including snake, scorpion, dog, rat, beetle, bug, etc.), sexual diseases (including gonorrhea, syphilis. etc.), jaundice, rheumatism, skin diseases, ulcers, dysentery, diabetes, and common cold and fever.

The bioresource potential, especially the medicinal importance of TDEF species, deserves detailed documentation in the additional unstudied sites. Further researches for bioresource augmentation and full utilization include: (a) developing propagation and nursery techniques for large-scale multiplication of multi-beneficial species and species of high medicinal importance such as *Sansevieria roxburghiana* (used for ear diseases and cough, and yielding silky fiber, face cream from leaf mucilage, sand binder, and a hedge plant), *Amorphophallus sylvaticus* (for piles), etc.; (b) phyto-chemical screening and bioprospecting of important species such as *Memecylon umbellatum*, *Cassytha filiformis*, *Cissus vitiginea*, *Sarcostemmaacidum*, *Atalantia monophylla*, and *Jasminum angustifolium*.

Fig.: Selected medicinal plants from TDEFs. a. *Sansevieria roxburghiana* (Agavaceae)-endemic herb, medicinal & silky-fiber b. *Strychnos nux-vomica* (Loganiaceae)-seeds medicinal; c. *Calophyllum inophyllum* (Clusiaceae)- seed oil medicinal; d. Night blooming, fragrant-*Tarenna asiatica* (Rubiaceae); e. *Strychnos minor* (Loganiaceae)-Hook climber with foetid flowers; f. *Cassia auriculata* (Caesalpiniaceae)-leaves and flowers medicinal.

Conservation Significance

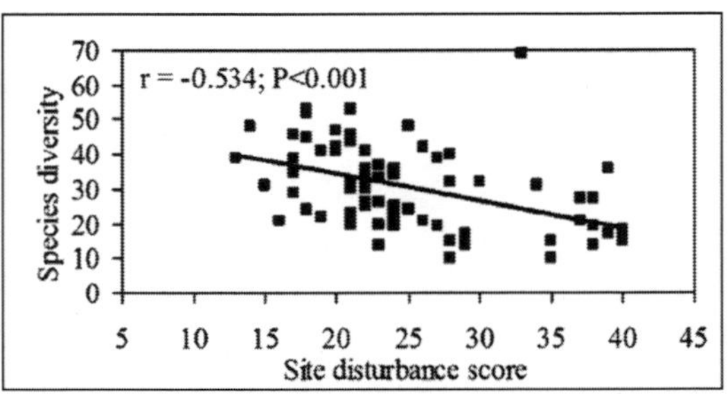

Fig.: Relation between site disturbances and species richness in 75 TDEF sites along the Coromandel coast of peninsular India

Overall disturbance scores of the 75 studied TDEF sites ranged from the lowest score of 13 forSuranviduthi to a maximum of 40 in Avudayarkoil and Azhagarkoil. Of the total 75 TDEF sites, 19 are relatively undisturbed (score range 13-20), but a moderate level of disturbance (score 21-30) is operative in 42 sites, whereas disturbance is severe (31-40) in 14 sites. The declining trend in mean species richness was observed from relatively undisturbed to highly disturbed sites and a significant negative correlation (r = -0.534; *P*<001) existed when species richness was plotted against site disturbance scores. In a linear regression analysis, when various site disturbance scores wereregressed with species richness, site disturbances such as resource removal, frequency of peoples' visit to a temple, and forest areal extent had a greater influence on species richness of the site.

Table. Conservation significance of TDEF sites ranked into three categories based on site disturbance scores, and illustrative site examples with their characteristic features.

Site disturbance	Area (ha)	Mean species richness	No. of sites	Illustrative site examples with characteristic features
Relatively undisturbed (= 20)	0.8-10	39.10 ±12.2	19	Araiyapatti - mono-dominant forest of *Strychnos nux-vomica* Suranviduthi - old growth mono-dominant forest of*Manilkara hexandra* Thirumanikuzhi – *Tricalysia sphaerocarpa* & *Lepisanthestetra phylla* dominated Shanmuganathapuram – mono-dominant forest of*Memecylon umbellatu m* Mavadipalayam – mono-dominant forest of *Garciniaspicata*
Moderately disturbed (21-30)	1-8	29.33 ±12.2	42	Arasarkulam - old growth, fragmented forest; dominants – *M. hexandra* & *Sapium insigne* Karukkai – culturally valued - temple; dominants - *Atalantiamonophylla* & *Pyrenacantha volubil is* Puthupet – high visitation, forest clearing in places for road & building construction Keeranoor – more openness, dominant - *Albizia amara* Kothattai – unique landscape in undulating one portion, old growth forest on sandy soil
Highly disturbed (31-40)	0.5-3.5	20.38 ±7.6	14	Kadapakkam – high level of resource extraction Azhiyanilai – converted native *Acacia leucophloea*plantation Keezhoor – forest converted to tank Embalam – converted to *Khaya senegalensis* plantation Avudayarkoil – converted into *Eucalyptus* plantation

Having changed in land use patterns and shrunken acreage (0.5 to 3.5 ha), the highly disturbed TDEF sites are largely converted to monoculture plantations. There are sites with least disturbance scores harboring the best natural dry evergreen vegetation with high diversity of plants of medicinal and cultural importance, and enhancing landscape heterogeneity as well as protecting the microclimates.

Moderately disturbed sites are currently exposed to high level of threats, primarily from humanintervention. Major issues like consistently increasing human habitation surrounding the forest area, poverty and illiteracy among large sections of the population, continuous areal shrinkage, over-exploitation, site degradation and land conversion are to be considered while assessing the conservation significance of each TDEF site.

Fortunately, the sacred grove status of TDEF sites largely helped to preserve the biodiversity along with the cultural values and religious taboos. People conserve the forests in undisturbed sites through a strict code of conduct on religious beliefs for several generations without any legal administration and clearly defined management policy and make only minimal resource extraction. Cultural transformations, eroding cultural values, and the changing world view of nature, especially among younger generations, has made this traditional forest management worse in many of the moderately and much disturbed sites.

In conclusion, the conservation of TDEF sites is important considering the restricted geographical distribution and representation of the unique and under-studied TDEF type, the extant level of biodiversity, and bioresource potential including medicinal plants and the socio-economic and ecological values of these systems. We recommend the following as long-term conservation strategies to preserve these sites:

(a) Promote awareness of biodiversity and bioresource values and cultural traditions associated with the sacred groves to people living around the TDEF sites—people who are also dependent on the forests and their resources and stand to benefit from conserving the sites that still remain relatively undisturbed;

(b) Restore moderately disturbed sites with characteristic TDEF species, involving the local communities in restoration programs and also in nurturing the planted saplings;

(c) Immediately protect and conserve much-disturbed sites by providing legal status to the forests and developing forest management systems involving the local community.

EAST DECCAN DRY EVERGREEN FORESTS

The East Deccan dry evergreen forests are an ecoregion of southeastern India. The ecoregion includes the coastal region behind the Coromandel Coast on the Bay of Bengal, between the Eastern Ghats and the sea. It covers eastern Tamil Nadu, part ofPuducherry and south eastern Andhra Pradesh.

Fig. Dry evergreen forests at Visakhapatnam,Andhra Pradesh

SETTING

The East Deccan dry evergreen forests cover lie in the rain shadow of the Western Ghats and Eastern Ghats, which block the rain-bearing summer southwest monsoon. The ecoregion covers an area of 25,500 square kilometers (9,800 sq mi), extending from Ramanathapuram District of Tamil Nadu to Nellore District of Andhra Pradesh. Much of the ecoregion is densely settled, and has been substantially altered by human activity, including agriculture, grazing, and forestry, over the centuries. The ecoregion is home to the metropolis of Chennai (Madras), and a number of other cities, including Pondicherry, Thanjavur, Kanchipuram and Nellore. It is estimated that 95 per cent of the original forest cover has been cleared, and the species composition of the remaining forests have been altered by intensive human use including the removal of all the taller trees.

Rainfall averages 800 mm/year, and mostly falls during the highly variable northeast monsoon between October and December. Unlike most of the world's tropical and subtropical dry broadleaf forests, whose trees tend to lose their leaves during the dry season to conserve moisture, the East Deccan dry evergreen forests retain their leaves year round. Only two other ecoregions exhibit a similar pattern, the Sri Lanka dry-zone dry evergreen forests and the Southeastern Indochina dry evergreen forests.

The ecoregion is home to two important wetlands, Kaliveli Lake in Viluppuram District of Tamil Nadu, and Pulicat Lake north of Chennai. Kaliveli Lake is one of the largest wetlands in peninsular India, and is deemed a wetland of national and international importance by the IUCN. It is a seasonal wetland, with a gradient from freshwater to brackish water, and is an important feeding and breeding ground on migratory bird flyway. It is currently threatened by encroachment by agricultural fields, wildlife poaching, loss of the surrounding forests, and increases in commercial prawn farming.

FLORA

The original vegetation of the ecoregion consisted of forests with an understory of evergreen trees and an emergent canopy of taller deciduous trees, including Sal (*Shorea robusta*), *Albizia amara* and *Chloroxylon* spp. Intensive human use of the forests over the centuries has mostly eliminated the deciduous canopy species, and the ecoregion's remaining forests are now characterized by areas of leathery-leaved evergreen forest, with a relatively low (10-meter) closed canopy. Predominant species are *Manilkara hexandra*, *Mimusops elengi*, Ceylon Ebony (*Diospyros ebenum*), Strychnine tree (*Strychnos nux-vomica*), *Eugenia* spp., *Drypetes sepiaria*, and *Flacourtia indica*. A few small enclaves of deciduous Sal forest exist, but are under intensive human pressure.

Only five percent of the ecoregion remains in forest, which is found in isolated pockets. Most of the ecoregion's forests have been degraded into tropical dry evergreen scrublands, characterized by thorny species such as *Ziziphus glaberrima*, *Dichrostachys cinerea*, *Catunaregam spinosa*, and *Carissa spinarum*.

FAUNA

Mammals found in this ecoregion include the Dhole (*Cuon alpinus*), Sloth Bear (*Ursus ursinus*) and Indian Spotted Chevrotain (*Moschiola indica*).

PROTECTED AREAS

Less than one percent of the ecoregion lies in reserves or protected areas but many are very small such as the sacred grove near Marakkanam in Tamil Nadu, which preserves a section of evergreen closed canopy forest. Several other temple groves in the surrounding area, including Puthupet, Pillaichavadi, Mudaliarchavadi, and Kottakarai, preserve small enclaves of forest. The Point Calimere Wildlife and Bird Sanctuary protects a 17.26 km^2 enclave of dry evergreen forest, as well as tidal wetlands and mangroves. Other preserves in the region include Vettangudi Bird Sanctuary (30 km^2) in Sivaganga District of Tamil Nadu, and Nelapattu Bird Sanctuary (160 km^2) on Pulicat Lake in NelloreDistrict of Andhra Pradesh. Srivilliputhur and Manjampatti Valley are refuges for the threatened grizzled giant squirrel (*Ratufa macroura*).

TROPICAL FORESTS OF INDIA: NOTES ON DRY TROPICAL FORESTS OF INDIA !

TROPICAL DRY EVERGREEN FORESTS:

Along the coasts of Tamil Nadu are areas which receive annual rainfall of about 100 cm mostly from the north-east monsoon winds in October-December.

Here, the mean annual temperature is about 28°C and the mean humidity is about 75 per cent. These areas are covered by the tropical dry evergreen forests.

The growth of evergreen forests in areas of such low rainfall arouses great botanical interest, and the reason for such a phenomena is difficult to explain. It may be due to the seasonal distribution of rainfall. Most of rainfall occurs in winter. The chief characteristics of these forests are short statured trees, upto 12 m high, with complete canopy, mostly of coriaceous leaved trees of short boles, no canopy layer differentiation, bamboos are rare or absent and grasses not conspicuous.

The important species are khirni Jamun, kokko, ritha, tamarind, neem, machkund, toddypalm, gamari canes, etc. Most of the land under these forests has been cleared for agriculture or casuarina plantations.

TROPICAL DRY DECIDUOUS FORESTS

These are similar to moist deciduous forests and shed their leaves in dry season. The major difference is that the species of dry deciduous forests can grow in areas of comparatively less rainfall of 100-150 cm per annum.

They represent a transitional type; on the wetter side, they give way to moist deciduous and on the drier side they degenerate into thorn forests. Such forests are characterised by closed and rather uneven canopy, composed of a mixture of a few species of deciduous trees, rising upto a height of 20 metres or so. Enough light reaches the ground to permit the growth of grass and climbers. Bamboos also grow but they are not luxuriant.

The tropical dry deciduous forests are widely distributed over a large area. They occur in an irregular wide strip running north-south from the foot of the Himalayas to Kanniyakumari except in Rajasthan, Western Ghats and West Bengal. The important species are teak, axlewood, tendu, bijasal, rosewood, amaltas, palas, haldu, kasi, bel, lendi, common bamboo, red sanders, anjair, harra, laurel, satin wood, papra, achar, sal, khair, ghont, etc.

Large tracts of this forest have been cleared for agricultural purposes and these forests have suffered from severe biotic factors such as over cutting, over grazing and fire, etc.

TROPICAL THORN FORESTS:

In areas of low rainfall (less than 75 cm), low humidity (less than 50 per cent) and high temperature (25°-30°C), there is not much scope for thick forests and only tropical thorn forests are found. The trees are low (6 to 10 metres maximum) and widely scattered.

Acacias are very prominent, widely and pretty evenly spaced. Euphorbias are also conspicuous. The Indian wild date is common, especially in damper depressions. Some grasses also grow in the rainy season.

These forests are found in the north-western parts of the country including Rajasthan, south-western Panjab, western Haryana, Kuchchh and neighbouring parts of Saurashtra. Here they degenerate into desert type in the Thar Desert. Such forests also grow on the leeside of the Western Ghats covering large areas of Maharashtra, Karnataka, Andhra Pradesh and Tamil Nadu.

The important species are khair, reunjha, neem, babul, thor, cactii, khejra, kanju, palas, ak, nirmali, dhaman, etc.

3

Joint Forest Management

TWO CHEERS FOR JOINT FOREST MANAGEMENT

From the inception of state forestry in India, perceptive critics have argued for a democratization of resource control, for a correction of the commercial bias promoted by successive governments, and for a proper participation in management and decision making by local user groups. Arguments first offered in the 1870s, and reiterated in subsequent decades, were revived, or reinvented, in the 1970s by the now-famous Chipko movement. It is no accident that Chipko originated in Garhwal and Kumaun, the part of India that has seen some of the most intensive conflicts between the state and the peasantry over forest resources.

The 1970s were marked by a series of forest movements in different parts of India. These took place in the Himalaya, in the Western Ghats, and, above all, in the vast tribal belt extending across the heart of peninsular India. In the Chotanagpur plateau, forest protests formed an integral part of the larger movement for a separate tribal homeland of Jharkhand, carved out from the huge, unwieldy, and predominantly non-tribal state of Bihar. In one much celebrated case, tribals demolished a plantation of teak, a highly prized furniture wood, that was coming up on land previously under the sal tree (Shorea robusta), a species of far greater benefit to the local economy.

Their slogan, "Sal means Jharkhand, sagwan (teak) means Bihar" was a one-sentence critique of the narrow commercial ends of state forestry. Since the 1970s, there has been an ongoing, nationwide debate on forest policy in India, a debate fuelled by the continuing social tension in forest areas and the evidence of massive deforestation provided by satellite imagery.

This debate has passed through three distinct if chronologically overlapping phases. The first phase might be designated the "politics of blame." The activists speaking on behalf of disadvantaged groups have held the forest officials responsible for environmental degradation and popular discontent. The officials, in turn, have insisted that growing human and cattle populations are the prime reason why fully half of the 23 percent of India legally designated as "forest" was without tree cover. The forestry debate of the 1970s and the 1980s drew, at times, on the heritage of earlier movements and critiques. The peasants of Garhwal and Kumaun, as this writer found out while doing field work there in 1982-83, were acutely conscious of how Chipko itself drew on a long and honorable history of peasant resistance to state forestry.

Tribal activists in Madhya Pradesh and Bihar, meanwhile, were not unfamiliar with the work and message of Verrier Elwin. And in the villages of the Deccan, social workers liked to offer the same quote of Jotirau Phule's reproduced earlier in the chapter, as proof that in the agro-pastoral system of that region, proper access to forests and pasture was vital to survival, and that it was the "great superstructure" of the Forest Department that continued to deny herders and farmers this access.

However, perhaps the most direct connection between the past and the present of forest management was effected in the summer of 1982, when the Government of India circulated a new draft forest act. Activists and academics joined hands to demonstrate how the proposed legislation was solidly based upon and, indeed, took further forward the centralizing thrust and punitive orientation of the notorious Indian Forest Act of 1878. After a countrywide campaign, the draft bill was finally dropped by the state.

As tempers cooled and polemic exhausted itself, a second phase, the "politics of negotiation," originated. In villages and state capitals, forest officers and their critics found themselves at the same table, talking and beginning to appreciate, if not fully understand, the other's point of view. Concessions were made by each side, protests suspended by one, and leases of forest produce to industry cancelled by the other.

One product of the growing dialogue between activists and bureaucrats was the approval, by the Indian Parliament in 1988, of a new National Forest Policy. Where the ruling Forest Policy of 1952 had stressed state control and industrial exploitation, the new document instead emphasized the imperatives of ecological stability and peoples' needs.

Then, slowly and hesitatingly, commenced the third phase, "the politics of collaboration." In the state of West Bengal, for example, the Forest Department initiated remedial action on its own, abandoning its traditional custodial approach by inviting peasants to cooperate with it. Thousands of village forest protection committees were constituted, each of which pledged to protect nearby forests in collaboration with the state. Thus previously authoritarian government officials joined with previously suspicious villagers to succesfully regenerate the degraded sal forests of southwestern Bengal. The success of "Joint Forest Management" or JFM in West Bengal has encouraged scholars, activists, and sympathetic civil servants to demand its replication in other parts of India. Outside its original home, however, the progress of JFM has been slow. Administrative styles and cultures of governance vary widely among the states and regions of India. So do individual orientations, with some forest officials still loathe to relinquish control, while others have been inspired to start village protection committees on their own.

A mapping of the forestry debate in contemporary India would therefore show significant regional variations. Some states are still stuck in the "politics of blame"; others have moved tentatively to the "politics of negotiation." West Bengal and parts of Andhra Pradesh, Madhya Pradesh, and Himachal Pradesh have instituted the "politics of collaboration" through the creation of JFM regimes. In this last scenario there is abundant scope for improvement. As analysts have shown, the JFM model now promoted by the Government of India reflects and sometimes reinforces inequities within rural society.

Gender and caste are two axes of discrimination, with women and low-caste members of the village community not having adequate representation or voice in the decision-making process (this is also true, to a great extent, of the van panchayats in Kumaun.) Likewise, pastoral groups and artisans, who have legitimate claims on forest resources, are sometimes given short shrift. Moreover, the forest officials still claim a monopoly of "scientific expertise," refusing to entertain villagers' own ideas on species choice, spacing, or harvesting techniques.

One serious problem with the JFM model, as currently promoted by the state and donor agencies, is that it allows the constitution of village forest committees only on forestland with less than 40 percent crown cover. This is a deeply constricting rule, which reserves to the state, and the state alone, exclusive rights over the bestclothed lands of India.

Thus forests situated close to hamlets cannot come under JFM regimes if they have more than 40 percent tree cover. Again, the regulations, strictly interpreted, would mean that if local communities were to effectively protect and replenish degraded lands, such that the crown cover was to come to exceed that magic figure of 40 percent, the state could step in and remove the area from JFM-which would be a bizarre outcome indeed.

Nor have changes in policy and orientation been accompanied by concomitant changes in legislation. Thus, the present regime is not flexible enough to allow for spontaneous community-initiated forest regimes to exist along with more orthodox JFM regimes. In some parts of India, the Forest Department is casting a covetous eye on areas well protected by village communities.

Thus in the Uttar Pradesh hills, the old established panchayat forests, managed by villagers, are sought to be brought under the JFM system only so that bureaucrats would have a greater say in their management. A new, carefully thought out Indian Forest Act is called for, which allows both for areas to be managed under state-village partnerships as well as by self-generated, autonomous community regimes.

One can thus envision a fourth (and possibly final) phase for the Indian forestry debate, the "politics of partnership." For collaboration, even where it does exist, takes place on terms set down by the state, through the officials of the Forest Department. We need to move on to a more inclusively democratic structure, where the state listens to and learns from the community, and where the community itself recognizes and deals fairly with the inequities within its own ranks.

The evidence suggests that contemporary advocates of decentralized forestry have had far greater success than their precursors. One reason for this is the altered political context: Brandis, Pant, and company worked under a colonial, authoritarian regime; the partisans of Chipko and similar movements in a democratic system. The revival of forest protest in the 1970s also coincided with the international environmental debate, which foregrounded the use and abuse of forests worldwide. The work of Indian scholars had, meanwhile, demonstrated with authority that the century-old history of state forestry in India must be reckoned a failure, in both an ecological and social sense. Finally, the problems with government-directed development programmes in much of the Third World had led to an increasing interest in nongovernmental forms of management and control.

These calls for forest reform from the outside were complemented by pressures from change from within. Starting with West Bengal, the governors themselves, namely the forest officials in charge of their vast landed estate, realized that old methods of control and exclusion were merely fuelling social conflict. An overworked and underfunded bureaucracy then started, slowly, to involve communities in forest working. What started as a strategic imperative became, at least for some forest officers, a sincere change of heart.

Once the critics from without were being echoed by the dissidents from within, the process of reform accelerated. This is indeed the signal lesson of Indian forest history-that meaningful policy change comes about only when the sustained pressure by social movements and their intellectual sympathizers resonates with the feelings of powerful officials within the state bureaucracy.

One or the other, by itself, will not do. When Brandis was active, he was handicapped both by his lone dissident voice within the Forest Department, and by the fact that there had not yet emerged an effective critique from outside. When Elwin, Mira Behn, and others propagated the feelings and aspirations of the peasants and tribals they worked with, the forest bureaucracy was, collectively and to the last man, deaf to their arguments and entreaties.

Forest policy remained unbending and unchanging, with the exception only of the Kumaun hills. There, as we have seen, the popular protests and outside critics were partially successful not because of a honest rethinking by the state but by its concern that this sensitive border region must not be tempted into outright rebellion. Elsewhere, where this political imperative did not come into play, the colonial regime refused to heed the widespread criticisms of its system of forest exploitation.

In more recent times, however, the radical critics have been aided by the autocritique of influential sections of the forest establishment. This confluence of external pressure and internal rethinking explains why, and how, the contemporary proponents of community forestry have, unlike their predecessors, been able to see their ideas and polemic become translated into official policy and (though less assuredly) into official practice. Nonetheless, there are indeed striking parallels between the ideas underlying the application of joint forest management today and the ideas of the early, prescient, and brave but for the most part unheard critics of state forestry discussed in this essay. With respect to the role of forest dependent communities, for example, there is a shared faith in indigenous knowledge, in the management capacity and robustness of local institutions, and above all, a sharp focus on local access to the usufruct of the forests.

Again, with respect to the role of the state, there is a common recognition of the essentially advisory role of the forest department, of its need to collaborate with rather than strictly regulate customary use, and of the justice of sharing revenues from forest working with the villagers.

Then, and now, critics have called strongly for an attitudinal change among state officials, a retraining and retooling in keeping with the democratic spirit of the age. Finally, both past and present proponents of decentralization seem to converge in their larger vision for forest policy in India, a vision which in my understanding consists of three central principles: (1) that benefit sharing (between state and community) and local control are to be the key incentives to ensure sustainable management and minimize conflict; (2) that community-controlled forests would work as a complement to a network of more strictly protected areas, further from habitations, that continue under more direct state control; (3) and finally, that the restricting of state control to these latter areas is vital on grounds of equity (*i.e.*, the respect for local rights and demands), efficiency (*i.e.*, as the most feasible course, with the state not biting off more than it can chew), and stability (*i.e.*, as the most likely way to lessen conflict).

There is little question that the ongoing attempts at reversing or mitigating state monopoly over forest ownership and management do constitute a significant departure from past trends. In a deeper sense, however, contemporary attempts at fostering participatory systems of forest management hark back to a much older tradition. In the late twentieth century, as in the late nineteenth century, there has arisen a movement for the democratization of forest management, for a system founded not on mutual antagonism but on genuine partnership between state and citizen. The first inspector general's vision for Indian forestry was abruptly cast aside in the 1860s and 1870s, but it may yet come to prevail. That would be a vindication of the life and work of

Dietrich Brandis, but also of Jotirau Phule, Verrier Elwin, Mira Behn, Govind Ballabh Pant, and the Poona Sarvajanik Sabha. Ramachandra Guha is an author and columnist based in Bangalore. His books include The Unquiet Woods, and Environmentalism: A Global History. He is now working on a history of independent India.

SOCIAL VALUES IN COMMERCIAL FOREST MANAGEMENT

Apart from the policy, legal and ethical reasons noted above, there are good commercial reasons to consider social values. Forest managers are increasingly subject to pressures from other interest groups, frequently concerning social values.

- Demands from forest peoples' groups may include greater respect for local populations' rights, carrying out or desisting from specific management practices, and support for their own environmental/social projects.
- Demands from local groups to contribute to social and economic development may include support for local enterprise, employment opportunities, excision of specific areas from management, and use of company infrastructure.
- Demands from trades unions and their initiatives may include greater respect for workers' rights to organize and negotiate, fair wages and benefits, health and safety at work, and the right to skills development throughout the organization.
- Demands to recognize other forest actors' rights to monitor, control and negotiate may include the development of agreements, and procedures to manage conflicts and make compensation.

Unless an active, organized approach is taken to respond to such pressures, forest managers may find themselves facing:

- Slow-downs, strikes, blockades, boycotts, legal battles;
- damage to forest stock;
- Arson, sabotage or vandalism to equipment and infrastructure;
- Reappropriation of forest lands for cultivation, or migration;
- Development of 'cultures of resistance' among forest-dependent people and their supporters (such as some consumers).

Considerable management skill and time may need to be invested in dealing with disputes, legal challenges and compensation claims, stalled negotiations, 'bad press' and political backlashes and general hostility towards forest managers and companies.

In contrast, participatory approaches that develop people's potential contributions can benefit forest managers, by:

- Improving the reputation of forest managers and companies;
- Uncovering and sharing useful local information;

- Broadening the base of ideas, skills and inputs applied to forestry;
- Efficient apportioning of responsibility, *e.g.* local groups may be better suited to managing recreation and harvesting of non-timber forest products;
- Better understanding of broader social, and thereby market, trends;
- improving transparency, accountability and therefore trust between all parties;
- Longer-run cost savings and risk reduction.

By taking an active approach to people as well as to trees, forest managers greatly increase the potential social benefits from forest management and their chances of support from others. Such experience should also increase their capacity to anticipate future developments regarding social values, for example in legislation or the market, and thus to gain 'first-mover' advantage from the situation.

If some social values are indicators of market trends (and they frequently are), then it behoves forestry organizations to keep close track of them. Many companies have made good business ventures by diversifying into recreation provision in particular. And it is increasingly clear that there is considerable financial value attached to brand names of certain leading companies which are known to produce social and environmental benefits alongside fibre. For example, the Greenpeace name has been estimated to be worth hundreds of millions of dollars.

CODES OF PRACTICE AND CERTIFICATION STANDARDS ON SOCIAL ISSUES

All initiatives to define SFM principles and criteria cover social values. Some of these are becoming enshrined in legislation, while others face forest enterprises through market relations.

Current systems of forest certification specify various social requirements at the level of the forest management unit. Even so, social issues remain perhaps the most contentious of certification standards, and there are differences between standards, particularly regarding the required degree of involvement of different interest groups in forestry and treatment of peoples' rights. There are also differences in interpretation of the standards by certifiers/assessors. Whilst some confine themselves to assessing the local social outcomes of forest management, others assess the roles of local stakeholders in the management of the forest and indeed the enterprise, and call for changes that appear to reflect their own biases. Whilst it is increasingly clear that forestry should not be a principal means of social engineering, and less still should forest certification, it is generally agreed that social standards for forestry will become more stringent in future. Note that some require precise performance thresholds to be met, while others require only that the issue shall be measured. Still others consider that the social issue is a policy concern and require it to be covered in policies only.

PARTICIPATORY FOREST MANAGEMENT

With the Pace of Population booming, increased energy consumption, over exploitation of the natural resources and rapid depletion of the forest reserves accelerated natural disaster like flood, drought and cyclones in Bangladesh.

Once Bangladesh was famous for its evergreen/ semi evergreen tropical and world famous mangrove forest. But over the years due to over exploitation of forests and its non-participatory management, more than 50% of the forest resources has been depleted. Realising the grim effect of destruction of forests and to repair the lapidated environmental condition, both the government and nongovernment organization have taken up afforestation programme. The NGOs have added a new dimension in the forest management, which has ensured participation of the community people and protection of the vegetation. Although, the government has also adopted participatory forest management but due to bureaucratic attitude easy access of the poor habitants are restricted in many cases. To overcome these situations, the existing government forestry policy, which was formulated in 1994, needs radical modification. There should be room to accommodate the NGOs, grass root organisations and general people in policy formulation, execution and evaluation of the programme.

Bangladesh lies in the North-Eastern part of South Asia between 20°34' and 26°38' North in latitude and between 88°1' and 92°41' East in longitude. The total area or the country is 144,000 sq. km with a population of about 120 million, density of population is 800 person per square kilometer. The most densely populated country in the world, Bangladesh is mainly a floodplain delta, which is formed at the confluence of the Ganges, the Brahmaputra and the Meglina rivers. Natural forest represents only 6 percent of the total land area of the country and is managed and controlled by the government. Village forest which contains annual and perennial trees and still provides a major source of food and income for majority of people, is managed by private individuals.

A rapidly increasing population is placing growing demand on natural resources, especially forest sector is under pressure to become more productive and efficient to keep pace with increasing demand. At present forest in Bangladesh is in unfavourable situation in terms of meeting increasing demand and also not adequate for maintaining ecological balance. This is primarily due to heavy population pressure and limited resource base, secondly, lack of integrated planning for development of multiple resource base with active participation of people resulting in high degree of environmental degradation, as illustrated mostly by deforestation and destruction of Natural resources.

In the phase of rapid depletion of forest aggravated by increasing demand for forest resources, and considering the prevailing socio-economic condition of the country, Government has put emphasis on participatory approach in development of forest resources of the country.

FOREST SITUATION IN BANGLADESH

Bangladesh has lost over 50% of its forest resource over the period of about 25 years. Actual forest coverage is only 6 percent of the total area and the situation is worsening despite of an attempt to preserve it. At approximately 0.02 ha per person of forest, Bangladesh currently has one of the lowest per capita forest ratio in the world. In Bangladesh, government-owned forest area covers 2.19 million ha, with the remaining 0.27 million ha being privately controlled homestead forests. Of the government owned forest land, 1.49 million ha are national forests under the control of the Department of Forest, with the rest being under control of local governments. Of the state owned forests, over 90% is concentrated in 12 districts in the Eastern and South-Western region of the country. However, due to over exploitation these forests have become seriously degraded. The natural forests of the country are classified into three categories: 1) Tropical evergreen/ semi-evergreen forest in the eastern districts of Sylhet, Chittagong, Chittagong Hill Tracts, and Cox's Bazaar: 2) Moist/dry deciduous forest also known as Sal forests in the central and the northwest region and 3) Tidal mangrove forest along the coast, known as the sundarban, the largest mangrove ecosystem in the world. These forests are official reserves and placed under the jurisdiction of the Forest Department. Unfortunately, recent inventories indicate a continuing depletion of all major forests.

Forest Management in Bangladesh

In Bangladesh management of government forest is the responsibility of the Forest Department under the Ministry of Environment and Forest. In this process the department is managing, protecting, developing the forest resources, forest land and also collecting the revenues. People have never been consulted nor involved in forestry activities. From the management point of view, forest of Bangladesh are being divided into three categories such as:

- State owned forest under the administrative control of Forest Department.
- State owned forest under the administrative control of Ministry of Land through District administration.
- Private village forest managed by private individuals. Forest under Forest Department control and management again divided into three major types viz; (a) Hill Forests; (b) Plain land Sal Forests, (c) Mangrove Forests.

Hill Forests: The tropical evergreen/semi evergreen forest cover as approximately 1.32 million ha of which 0.67 million ha is controlled by the forest department and rest is under the control of hill district council. Clear felling followed by replanting with suitable species (both long and short rotation) is the method of management in hill forest. Because of increased demand for timber and fuel wood and prevailing socio-economic condition of the country this forest has greatly affected and rate of denudation is considerably high. The forest department is mainly confined in raising of single species plantation. Inventory shows that most of these plantations would not give the desirable output. This programme suffers from technical, social and administrative soundness. Another problem is most of the high forest are subjected to shifting cultivation by the hill tribes. The tribes are entitled to shifting cultivation in forest land under administrative control of district administration which has resulted in the total destruction of these tropical evergreen forest. The growing stock has depleted from 23.8 million m^3 in 1964 to less than 20.7 million m^3 in 1998. Mangrove Forests: Known as Sundarbans, the largest mangrove ecosystem in the world. Sundarban forests are being managed by selection felling method followed by natural regeneration. Beside Sundarbans, plantations are being raised with mangrove species in the newly accreted char land all along the Coast of the Bay of Bengal. Sundarban forest is an official reserve forest, unfortunately recent inventory shows a continuous depletion due to over-cutting, illegal felling. It is estimated that in less then 25 years, the volume of commercial species Sundari, Gewa, has declined by 40 to 50% respectively.

Plain land Sal Forests: Silvicultural system applied for Sal forest was coppice with standard system. In this system matured trees were felled and the areas were protected for coppice regeneration. The typical nature of Sal forest is that this forest is scattered. In the forest areas there are agricultural lands owned by the adjacent people.

Frequently these land owners are extending their lands and encroaching to forest and in the process they are destroying the forest and subsequently converting the area to agricultural land. In this process forest lands are being marginalised day by day. FAO estimated that only 36% of the Sal forest cover remained in 1985; more recent estimates that only 10% of the forest cover remains due to over exploitation and illicit felling through there is an official base on logging since 1972. Most of the Sal forest are now substantially degraded and poorly stocked. The situation calls by for involvement of community people in the forest management.

Inventories how that there has been overall depletion in forest resources in all major state owned forest. The growing stock in Sundarban has been depleted from 20.3 million m^3 in 1960 to 10.9 million m^3 in 1998. In the Hill forest of hill districts, the growing stock has depleted from 23.8 million m^3 in 1964 to less then 20.7 million m^3 in 1998. Over-cutting by timber merchants, increased consumption linked to population growth, shifting cultivation, encroachment, illegal felling and land clearing for agriculture, lack of participatory management have been the principal causes of deforestation and shrinking of forest land in the country.

Since 1960 two major approaches regarding the role of forestry in development have been reflected in the forestry sector of Bangladesh. In the 1960's, Bangladesh as a part of Pakistan and then as an independent nation has followed 'An Industrialisation Approach' consonant with the international conventional wisdom at that time. As a result, Department of Forest raised large-scale Industrial plantation which were seen as conversion of low-yielding natural forest into artificial plantation of species (mostly teak) of great economic importance. This conversion of semi-evergreen and evergreen forest into deciduous teak plantation was largely concentrated in hill forest areas. During the plantation raising local people were not consulted and often they did not drive any benefits from these plantations. The lack of support by the local people/ communities in combination with lack of silvicultural knowledge and lack of proper maintenance contributed to raise low quality plantations and these plantations were also lost due to illegal felling. In the name of plantation the genetic resource of the ever-green/ semi-evergreen forest was lost. Forest Department was considered as revenue earning department. The main activities of Forest Department were concentrated in extraction of trees from the forest and replanting of those felled areas where applicable, Forest Department has not considered the people and their participation in managing forest of the country.

In the 1980s following a change in thinking about the role of forestry in development, and peoples participation in forestry activity was encouraged. People participation with the forestry sector realised the need of people oriented forestry programme to replenish the degraded forest resources of the country. Accordingly, in 1994 Government formulated a forest policy replacing earlier one enunciated in 1979 with a due emphasis to the need for people's participation in forest management.

PARTICIPATORY FOREST MANAGEMENT APPROACH

PAST ACTIVITIES

Forest extension activities were formally launched in the country in the year of 1964 with the establishment of two forest extension divisions at Dhaka and Rajshahi and later two divisions at Comilla and Jessore. It was really a very small programme and the activities were confined only to establish nursery in the districts headquarter and raised seedling and sell the same to individuals and organizations. The location of this programme was so urbanized and limited that it only partially served the needs of the effluent town dwellers only.

Betagi-pomora Comunity Forestry Project

The first community forestry programme in the country, started at Betagi and Pomora mouza (village) under the district of Chittagong in the year of 1979 with the personnal initiative of Prof. A. Alim, renowned forester and Prof. Dr. Mohammed Yunus, founder of Gramen Bank. Initially the project covered 160 ha of Government denuded hilly land at Betagi and with 83 landless participants from adjacent community and subsequently extend over another 205 ha of Government owned denuded hilly land at Pomora with another batch of 243 landless (families) participants. Under this programme each landless participant was provided with 1.62 ha of land for growing tree and hoticultural crops with technical and financial assistance from the Forest Department. This community programme has given the landless an identity of their own and a sense of direction in life. But this model has not been replicated in the other areas due to lack of initiative of the Forest Department as well as the Government.

Rehabilitation of Jhumia Families

Another project was undertaken by the Forest Department in the Hill tract areas to establish plantation through rehabilitation of Jhumia families in 1980. Main objectives of the programme were (i) to rehabilitate tribal families in the Unclassed State Forest(USF) lands along with rehabilitation of denuded USP land; (ii) to introduce a sustainable agroforestry production system; (iii) to improve the socio-economic condition of the tribal people and (iv) to motivate tribal people in development of forestry.

Under this programme each family was allocated 2.02 ha of USF land for growing agricultural crops (over 1.20 ha), raising plantation (0.80 ha) and for house construction (0.20 ha). The rehabilitated families were given land use rights and were allowed to enjoy 100% benefits accrued to those lands.

The participants were given input support for growing agriculture, horticulture and forestry crops and cash support for house construction. This programme continues for quite a long period of time but could not sustain mainly because of nomadic character of the tribal groups.

Another reason of failure was that the families were rehabilitated in clustered villages without considering their cultural and religious values. Thus in most of the cases it was found that the families have left the area.

A parallel programme was also initiated by the Chittagong Hill Tract Development Board in which Forest Department was responsible for implementation of afforestation component where Cittagong Hill Tract Development Board was responsible for the rehabilitation component. This programme was also not found so much responsive to hilly people except for some plantation establishment.

Development of Community Forests Project

The activities of the first phase of this project began in 1981 and were completed in 1987 in seven greater districts of the North-Western zone of the country.

The main components of the project were:

- Strip plantations along roads and highways, railways, canal sides, district and Union Parishad roads, totalling about 4,000 km.
- Fuelwood plantation on 4800 ha of depleted Government land on participatory concept.
- Agroforestry demonstration farms over 120 ha also with participatory concept.
- Replenishment of depleted homestead wood lots in 4,650 villages.
- Training of Forest Department Personnel and Village leaders.

Development of Forest Extension Services (1980-1987)

Development of Forest Extension Services (Phase II) began in 1980 with the Government funding and subsequently amalgamated in some areas (*i.e.* North-North West district) with Asian Development Bank funded Community Forestry Project. The main activities under this programme were:

- Afforestation in some 3100 villages.
- Roadside tree planting along 3600 km of primary highways and roads and about 600 km of Union Parishad roads.
- Production of 49 million seedlings for distribution.

Thana Afforestation and Nursery Development Project

This project is a follow-up of Development of Community Forestry Project and Forest Extension Project and has been designed primarily to: (i) increase the production of biomass fuels and (ii) enhance the institutional capability of FD and local administration in implementing a self-sustaining nationwide social forestry programme. In order to increase the production of biomass fuel and to arrest the depletion of tree resources, the project envisaged to develop tree resources base through planting of depleted sal forest as well as brining all suitable and available land in the rural areas under tree cover with active participation of the rural poor of the locality.

Originally the project was to be implemented by the Forest Department and former Thana Parishad during the period of 1987 to 1994. But in 1992 Government decided that the all project activities were to be implemented by Forest department alone. The major components of the project were:

1. Establishment of plantation over 20,225 ha depleted Sal forest areas.
2. Development of agroforestry over 4,200 ha in the Sal forest lands.
3. Raising strip plantation on 17,272 km along Road and highway, Railways, Embankment and Feeder Roads.
4. Raising 1,282 ha plantation in the land outside the BWDB.
5. Planting 7.017 million seedlings at the premises of different education, religious and social institutions
6. Establishment of 345 nurseries at Thana headquarters.

7. Raising of 10.618 million seedling for distribution to public.
8. Training of some 76,000 people of different levels.

Here this may be mentioned that at the last stage of the project implementation, the Government has found that this was quite impossible to protect the strip plantation and also impossible to trained 76,000 people by the Forest Department alone. The Government invited NGOs to participate in this programme for successful implementation. PROSHIKA, POUSH, GRAMMEN BANK and other NGOs came forward to help the Government for successful completion of the project; NGOs employed their group members to protect the strip plantation and ADAB came forward to train people at different levels with the help of its member organisations.

The above plantation activities were carried out with the direct participation of the local people with the help of the NGOs by executing benefit sharing agreement.

Coastal Greenbelt Project

Another project financed by Asian Development Bank is under implementation in the Coastal region of Bangladesh. The main objective of the project is to create a vegetative belt all along the coast to save the lives and properties of the people living in the coastal areas from devastated cyclone and tidal surges which occur very frequently in those areas. All of the activities of this project are also being carried out following participatory approach. In this project also the participants have been selected among the poor people living in the adjacent areas by involving NGO and a pre-designed benefit sharing agreements also being executed with the participants to protect their rights over plantations and to ensure benefit expected to be received out of the plantation.

Agroforestry Research Project

Pilot Agroforestry Research and Demonstration was implemented by the FD in the Sal forest areas. The project had been developed precisely to design/ develop agroforestry modules which is environmentally feasible, socio-economically acceptable enhance tree and crop production at the same time to uplift the socio-economic condition of the participants. The project aimed at using 120 ha of encroached Sal Forest land of Dhaka, Mymenshing and Tangail Forest Division to develop suitable participatory plantation models.

Food Assisted Social Forestry Programme

The World Food Programme assisted the Government to develop Social Forestry as a national programme and the Government incorporated WFP assisted social forestry programme in its annual development plan from 1998. Poverty alleviation, economic rehabilitation of rural poor especially the destitute women of the society by engaging them in forestry activities, social uplift of rural poor and environmental improvement are the main objectives of this project. Historically this programme was conceived in the country since 1989 on pilot basis allocating in kind resources (Wheat) to a limited number of NGOs for raising strip plantation along roads, embankments, Highways etc. in rural areas following the participatory mechanism. In implementing this programme FD was involved later on to provide technical guidance to the NGOs and other GOB agencies. At present probably this is the largest Participatory Forestry Programme in Bangladesh. From 1990, 100 NGOs are involved in this programme and at present about 60 NGOs are continuing with the programme. Commencing from 1990 up to 1998 about 31 million trees were planted involving 0.062 million people directly and 0.62 million people indirectly. The programme has created employment to the tune of 68 million man days. This programme is being implemented by the NGOs through contractual benefit sharing among participating poor men & women 60%, NGOs 10%, the rest land owners.

CLIMATE CHANGE MITIGATION AND ADAPTATION IN FOREST MANAGEMENT

The potential effects and significance of climate change on the forest sector. These impacts have varying consequences and are dealt with differently by forest managers. The possible operational options available to forest managers for addressing climate change mitigation and adaptation are assessed. In addition, the extent to which these options are being applied by forest managers is discussed, with the help of case studies. These examples, although not necessarily due to climate change, give good indications of what managers perceive to be good solutions to potential future changing climatic conditions.

FOREST MONITORING

Forest monitoring is very useful to detect changes due to climate change, natural disturbances or human activities. It has become a requisite in the context of climate change mitigation in particular, in relation to deforestation and forest degradation. Due to the potential benefits that accurate carbon monitoring may bring within the REDD+ framework, monitoring has developed greatly over the past few years and requisites on accuracy and acceptability have increased.

For monitoring in general to be successful, it needs to have clear objectives, be as simple as possible, and benefit the people that invest time and/or money in it. However, many times the objectives may be clear, but the activities that are needed to meet those objectives may be vague. This may be due to lack of experience or lack of certainty on how climate will change and how this possible change will affect different components of the forest and forest management. A tendency exists to want to monitor everything that might possibly change, resulting in impractical and expensive monitoring proposals. As a result, monitoring for the impacts of climate change on the forests and people related to the forest is still just emerging.

Monitoring helps us to identify changes and evaluate tendencies. Monitoring does not necessarily tell us the reason for these changes and tendencies, unless previous research has established such causal links. The next step would therefore be to analyse whether such changes correspond to changes in climate characteristics and then, to analyse whether such tendencies are negative or positive for the forest and the forest managers and whether actions can be taken to reduce the negative consequences and increase the positive ones. Current discussions on the implementation of REDD+ are occurring at the national level, however most of the monitoring experience has been obtained at the forest management unit level. While monitoring needs at these levels differ, they are highly complementary and any carbon monitoring system should consider linking these levels. It is very important to include all stakeholders to ensure agreement on the methodology and the variables to be monitored. The involvement of local actors has been shown to have two advantages; it is cheaper and creates greater ownership of the monitoring results.

In spite of the importance of monitoring for SFM and for preparation of responses to climate change, it still is not a common practice. Particularly in developing countries, few forest managers have the resources (human and financial) to implement these assessments.

Monitoring of Changes

Adaptation of forests requires in the first instance the identification of the changes that may occur and to which adaptation may be necessary or desirable. Although in general terms forest change scenarios can be developed based on global and regional climate change projections, the exact changes that will occur are not well known. There are several reasons for this uncertainty; the uncertainty in the climate change models in general, the scale at which climate change projections are made, the 12 inherent adaptive capacity of species and the communities they are in and the effect that interactions between species may have on adaptive capacity. In some areas, the changes that have been projected are drastic. The northeastern Amazon, for example, may lose most of its forest cover because of massive forest dieback due to droughts, giving rise to savannah vegetation. However, the rate of change and the exact result is not that clear. Other areas may follow suit at different rates and with different results. It will be difficult for forest managers to react to these changes, especially if it is not clear when and how these changes occur.

Adaptation strategies will need to include monitoring systems on climate, vegetation, fauna and essential non-biological components of the forests such as water availability. Without such monitoring systems, the forest manager will be grappling in the dark when making management decisions. In forestry, such monitoring systems are important, particularly because of the long time lapse between management actions and forest response. For this reason, permanent sample plots (PSP) are an integral part of SFM. Their main contribution to SFM has been a better understanding of the dynamics of forests and plantations.

PSPs have been used for stock-taking (both at a national scale and in continuous forest inventories), for monitoring of changes in managed and unmanaged forests (*e.g.* in certified forests to monitor changes in species composition and structure), and for research purposes (*e.g.* the effect of silvicultural treatments and harvesting on species composition, structure and biodiversity). PSPs are less useful to measure changes in the diversity of fauna, impacts of forest operations such as harvesting and impacts on ecosystem services that go beyond the forest plot boundaries (for example water flow).

In countries that have long-standing experience with PSPs and a good network of meteorological stations, PSPs may provide a good contribution to the analysis of the effects of climate change on forests. While PSP are good instruments to detect changes at the stand level, forests are also influenced by changes that occur on a landscape level, *e.g.* water quality affected by sedimentation. To detect such changes, a combination of remote sensing techniques and a network of PSPs is probably the most appropriate strategy: remote sensing to detect changes in forest areas, and PSPs to detect changes in forest quality.

Since remote sensing images and their interpretation for forest management is relatively costly for the forest manager, such monitoring is best carried out by organisations or associations that are responsible for larger areas or a group of stakeholders. This will require, that all potential users of the monitoring information agree on a common set of variables that are useful for forest management decisions and should therefore be monitored. An important part of monitoring systems is the database and processing of the data. This usually requires major investments in human resources but some companies have been able to develop their own computer hard and software that allows for quick data storage and analysis.

Monitoring Techniques of Animals

The techniques used monitor animal populations, particularly the larger mammals, depend on the objective of sampling. More local research is necessary to identify the techniques and variables to be sampled in specific cases (*e.g.* for a particular species in a defined region). Climate change can shorten the life cycle of insects, increasing their reproduction rate and the risk of infection and damage. Traps in sampling points can help to detect rapid increases in population sizes. The traps will need to be specific for the insects to be monitored, have appropriate bait and be placed at the right position in the forest. Turchin and

Odendaal (1996), found that one funnel trap, used to trap southern pine beetles in the united States, were good for covering an approximate area of 0.1 ha. Some insects are more ground related (*e.g.* dung beetles) while others (*e.g.* butterfly families) may fly in open or closed forest areas.

Although these latter insect groups have not been related to pests, they have been successfully used to identify changes in forest structure and composition related to fragmentation and tree harvesting, and may be useful to detect forest changes due to climate change. Specific dung beetles may be related to specific mammals and butterflies have been related to openness of the forest and may be an indication of dieback. Further research is needed to fully understand these relationships. Larger animals may be trapped (as in the case of small rodents), or counted visually using walking transects. Animal tracks may also be used as an indication of the presence and abundance of species. However, care should be taken that sampling density is sufficient to formulate robust conclusions. Steele *et al.*, (1984) concluded that three repetitions of a 2 km transect was sufficient to determine species abundance, richness and diversity of large animals, but it was more difficult to estimate small mammal richness and diversity. In general, design of a monitoring system requires expert knowledge, but local communities can be trained as parataxonomists to implement the monitoring.

Due to the need for additional information on species behaviour and preferences, as well as the relatively high time investments needed for animal monitoring, it is important to identify, as early as possible, those animals (and plant species alike) that are more susceptible to climate variations. Abundant animals may be easier to monitor, but many of them may also be less susceptible to changes in climate and the environment. Usually the species with a small range and short generation time are more responsive.

Forest Fire Monitoring

Monitoring of forest fires contributes to our knowledge on the extent of deforestation and forest degradation. Such monitoring is traditionally done through patrolling forest areas and operating watchtowers. The development of remote sensing techniques has made it possible to come to ever more accurate and timely information on forest fires, above all in large uninhabited areas. Laneve *et al.*, (2006) estimate that if images can be obtained at a sufficient spatial resolution to detect 1500 m^2 fires at 30 minute time intervals, this will be sufficient to reduce the number of large fires in the Mediterranean forest of Italy. For small forest land holders and many communities, such technology is not available and even the construction of towers may be too high an investment. Patrolling, however, has shown to be an effective way of forest fire prevention in community forests in Guatemala. During the last decade several proposals have been made to set up fire detection systems using wireless sensors, but most of these have not emerged from the experimental phase, possibly due to costs and the problem of maintaining the network.

CAPACITY OF FORESTS TO RESPOND TO CLIMATE CHANGE

The adaptive capacity of forests, for the purpose of this document, is understood to be the inherent ability of the forest to adjust to changing conditions, moderating harms and taking advantage of opportunities (Locatelli *et al.*, 2010). Thus, strengthening of the adaptive capacity is oriented at increasing the resistance or resilience to changes but may also include adapting the forest to new conditions by facilitating changes in the system (*e.g.* by species introduction). In general, strengthening the adaptive capacity of forests aims to maintain, restore or enhance forest area, biodiversity and forest health and vitality.

Many of the actions oriented towards mitigation of climate change through REDD+ have a strong potential for synergies with actions oriented at strengthening the adaptive capacity of forests, in particular if such actions consider ecological safeguards, such as biodiversity conservation. Experiences in strengthening the adaptive capacity of forests to climate change have been more widespread in plantations and agroforestry systems. These systems tend to have a simpler structure and composition that makes it easier to detect changes due to climate change and to design and implement adaptation-strengthening mechanisms. This is much more difficult in complex natural forests, in particular in the tropics. However, because of their simplicity, these systems may also be more vulnerable and therefore the need to look at adaptation options is greater. Interestingly, several of these adaptation activities are oriented towards making these (agro) ecosystems more diverse.

Maintaining Forest Area

Larger forests usually have greater species diversity and cover a greater variety of sites, thus reducing the risk of losing the whole system if climate change negatively affects several species or specific site conditions. Forest management appears to be the solution; both wellmanaged protected areas and well-managed community and private forest concessions in Guatemala and the South of Mexico have shown to be more effective in avoiding deforestation, fires and forest degradation than areas poorly managed areas. This probably goes beyond purely economic considerations.

Land and forest tenure, recognising the benefit of maintaining forests and joining forces with other forest managers are all important requisites. Size of forest area also seems to be important, for both individual landowners and community or multiple owners. Ecologically, larger forest areas show less edge effects, while from the management point of view, larger sizes allow for economies of scale.

Managing natural forests often is recognised as a claim on that forest. If good relations are held with local people, such claims are well respected. Owning forest but not managing it, has often resulted in unauthorised entry by third parties for the extraction of timber and NWFPs, or for conversion to agricultural land. Managing the forest but not entertaining good relations with the neighbours has often resulted in forest use conflicts, at times ending up in armed conflicts or burning of parts of the forest estate.

Such relations are more important in large forest tracks, since in these it is harder to establish continuous human occupancy. Good forest management normally includes fire, pest and disease management. Of these, fire management may be the most significant in maintaining the forest area, although serious pests, such as the mountain pine beetle in pine forests in North America, may also contribute to substantial forest loss. Managing the forest may be costly and income from the sale of one or more of its products may not off-set the extra cost of management. However, often, costbenefit analyses compare conventional operations (without much strategic planning or considerations for biodiversity or forest dependent communities) with managed operations.

From a private forest owner's point of view, this may be reasonable, but in practice, this approach has been used to justify continued conventional harvesting operations, giving the forest sector a poor image and increasing the pressure on governments to impose stricter regulations. In countries with greater willingness of the private sector to participate in 15 improving forest management, a series of alternatives were found to either make forest management attractive or propose other forest-based income solutions. Usually this was done by a carrot and stick approach: if a forest manager does better than the legislation requires, they receive subsidies, discounts on taxes and are a preferred provider of specific ecosystem services. While these approaches in theory seem to be very promising, in practice they have not had the expected outcomes. This is partially because of the high financial and administrative cost to actually obtain the carrots and because forest owners and managers were not aware of the existing opportunities.

Management of forests, therefore, should go beyond the mere planning of protective or productive activities and not depend on one single form of financial income. Forest managers need to be informed and aware of local, national and international opportunities for income generation. New opportunities may be through tax or fee discounts for good forest management (as the case of harvesting fee discounts for certified forests in Peru), payment for environmental services (*e.g.* Costa Rica and Mexico) or niche markets (*e.g.* markets for specific NWFPs and carbon).

Due to the relatively limited demand of these above mentioned products and services and with only a few exceptions (*e.g.* Brazil nut gathering in Bolivia, Stoian, 2004), none of these have been able to single-handedly pay for management and protection of large natural forest tracts. Only where combinations of products and services were obtained has management of natural forests become a serious land use competitor. New opportunities may also work for forest plantations, where payment for environmental services, such as carbon sequestration, may at least partially off-set the initial establishment costs.

However, selling sequestered carbon at the end of the rotation has little effect on the overall profitability of plantations, due to the low carbon price and high interest rates. The present value of these future sales is very low and rarely will change a non-profitable exercise into a profitable one. Selling less carbon earlier on during the rotation, or shortening rotation length, are two options that may make timber plantations more attractive. Establishing tree plantations not for timber, but only for carbon or other environmental services, as yet has to show its profitability for the forest owner or manager and is usually only accomplished where the forest owner or manager also garners important non-tangible benefits from those forest.

Conserving Biodiversity

Maintaining forest area is of course a good means to maintain a certain level of biodiversity. However, as can be seen from some countries, increasing forest area does not necessarily increase biodiversity nor does it necessarily mean that old forests or undisturbed forests are maintained. In many countries, reduction of net deforestation figures is at least partially due to compensatory measures, such as natural regeneration and plantations. Although depending on how these new forests are being managed, and how close they are (in time and space) to the original natural forests, these usually do not have the same species composition and biodiversity as the lost natural forests. In terms of capacity to adapt to climate change, the change in species composition may sometimes be an advantage, if new species are better adapted to changing conditions. More problematic may be loss of diversity. Loss of diversity will make forests more vulnerable to changes, since they will not have the rich gene and species pool from which to select for the new conditions. In this respect, care should be taken of the trade-offs between mitigation and adaptation objectives; too great an emphasis on management for carbon may reduce structural and compositional diversity, thus reducing the system's inherent adaptive capacity.

A noted change in forest management in the light of climate change has therefore been an increased interest in maintaining or increasing diversity of the forests. Mixed species plantations, use of a larger number of clones and reductions in the scale of harvesting operations have been implemented as measures to maintain or increase biological diversity. These same measures are now receiving more attention because of their potential benefit in preparing forests for climate change. In addition, literature suggests that the potential yield increase from 16 appropriately selected species mixes will more than outweigh the additional costs that may be involved in mixed tree plantation establishment. The use of nitrogen fixing tree species as part of the mix, in particular in degraded lands, is beneficial for overall growth rates.Reducing the scale of harvesting operations is one way of increasing the possibility of ecological connectivity between forest patches. Plantation establishment is also an important measure that may achieve this as does planting of trees outside the forest.

Although some practices are being adopted and theoretically will contribute to maintaining biodiversity, there is still a need for further research. For example, we do not yet know how much biodiversity change will cause a major and irreversible change of forest types or even ecosystems. Some authors suggest that maintaining functional diversity may be sufficient. However, it is unclear how susceptible diversity will be to climate change, if, for the different functions, the number of species that provide diversity function is strongly reduced. It is also unclear how species will react to climate change, by themselves, in combination with other species and in combination with a number of environmental and human factors. Continuous monitoring of the forests is critical to providing insight into these interactions.

Maintaining Forest Health and Vitality

The main threats to health and vitality are pests, diseases, fires and extreme weather events. In addition, diversity usually strengthens health and vitality and therefore the actions mentioned above to maintain or enhance biodiversity, also are useful for maintaining health and vitality. A number of silvicultural techniques have been developed for maintaining health and vigour of a stand. Removing old, poorly formed and damaged trees, for example, reduces the risk of spreading diseases and pests, although at the same time it may reduce diversity and thus increase the susceptibility of forests to diseases and pests.

Applying such treatments requires knowledge of the specific risks related to individual tree species and the potential benefits of maintaining poorly formed trees in the forest. Using harvesting residues on the forest floor may increase availability of nutrients for the remaining trees, thus increasing vigour, but may add to the fuel load and therefore increase fire risk. Nevertheless, timing of use (beginning of wet season) may increase the benefits and reduce the risks. In plantation forests, a reduction of old growth and an increase in the relative presence of young stands, enhances the general health and vigour of the forest from the point of view of timber production in the medium term. Again, however, it reduces diversity, thereby reducing the adaptive capacity of the forest to externally driven changes. The decision to reduce such growth in favour of young stands needs to be taken in consideration with local conditions and management objectives.

Reducing Risk and Intensity of Damage

Reducing the risk and intensity of pests, diseases, fires and hurricane damage, along with managing the hydrological cycle, will become major concerns for many, if not all, forest managers under changing climatic conditions. Due to the complexity of measures that may have contradictory effects, there is the tendency for integrated management practices; for example, combining insect control with monitoring exercises and implementation of management practices that reduce susceptibility of the forest to insect attacks. Such practices include those treatments that help maintain the vitality of the forest, including timely thinning and species mix. For most regions however, few comprehensive management plans exist, and in most cases, plans emphasise monitoring and combating pests and diseases, rather than preventing them.

The example of pine beetle management in Central America shows how management oriented towards the prevention of fires may also reduce the risk of insect pests. Integrated fire management in forests has been promoted in many countries and by international and national agencies alike. The proceedings of a 2009 seminar in California summarise many of the experiences and achievements to date. Martell (2007) defines forest fire management as: "getting the right amount of fire to the right place at the right time at the right cost". Integrated fire management includes the following components; prediction of fire occurrence, fire prevention, fire detection, initial attack, fire management, strategic planning of resources, and fuel management. Training, knowledge sharing and planning of the different components of fire management is extremely important.

Prediction of fire occurrence requires the identification of the main causes for forest fires, followed by an analysis of factors that influence the frequency of those causes. In Mexico, this has been done through proxy indicators, such as nearness of populations and types of land use (danger assessment). This needs to be combined with characteristics of the fuel, temperatures and moisture content (risk assessment) as well as with an assessment of the value of the resource in order to be able to set priorities in fire prevention and combat activities. In many developed countries, this information is combined with daily weather forecasts to assess fire danger ratings. This has proved to be a useful early warning system.

Fire prevention is probably the most cost effective and efficient way to reduce fire damage. It requires identification of frequency, size, severity and causes of human induced fires and needs to be followed by awareness campaigns to reduce those causes or projects oriented at training and introduction of alternatives to burning forest and grasing lands. In many regions, for example Mediterranean countries and western states of the United States, watch towers are the main tool for fire detection. Patrol flights are used as well but it is difficult to plan them cost effectively. Once detected communication of location, size and burning characteristics of the fires among the fire fighters and to the public is very important, especially for the initial attack.

Whereas in North America it is a problem of deciding how many air tankers to keep on standby for the initial response to fires, in many tropical countries air tankers are not available and initial attack may need to be done by local fire brigades. The quicker the response time, the smaller will be the risk of uncontrolled fires.

However, such fire brigades require an extensive programme of training of the local teams, as well as sufficient equipment to be able to fight small fires. Initial attacks become more efficient if these are based on previously prepared plans that consider potential impact of fire fighting, as well as the potential risks of damage to natural, human and physical resources.

Fire management requires continuous monitoring of fuel conditions, fire behaviour and climate. With this information and short-term predictions on their changes, fire fighters can decide when and how many people to dispatch to control the fire. Fuel management is one of the more common and discussed approaches to forest fire reduction. Protection of forests has led in many cases to the accumulation of fuel, increasing the risk of forest loss, rather than reducing it. Prescribed burning is a common tool, but this needs to be supervised by experts, in order to avoid the fires to get out of control. Cutting of fire breaks is another measure often taken. These do not only form barriers to fire progress, but may also help accessibility of the forest for fire fighting crews. Technology has proved vital in monitoring and control of forest fires in regions such as North America and Europe. However, in tropical countries, where access to this technology and to the forests is more limited, fire management strategies need to include all stakeholders and all related sectors since most fires originate outside forest lands and involve non-forest stakeholders.

Risk reduction strategies for damage from hurricanes or cyclones need to consider damage to forest, agricultural crops, communities and infrastructure. Reducing risk of damage to forests can be done through silvicultural practices such as shorter rotation cycles; young trees often are more resistant to wind throw, but if thrown, will result in less biomass lost compared to larger trees. Increasing resilience, for example by maintaining good seed sources and mixed species forests, including species that readily sprout after wind throw is another means of reducing damage. In Toncontín, Honduras, the local community started to improve forest management in 1997, one year before hurricane Mitch struck. The forest is dominated by species resilient to hurricanes through different strategies including rapid and abundant seed regeneration or re-sprouting. Part of the management improvement was the retention of seed trees. Comparative studies one year after Hurricane Mitch indicated that the forest with seed tree retention was recovering its original species composition faster than forests where all commercial trees had been cut without seed tree retention.

Improving Water Regulation

Changes in precipitation patterns probably will affect forests more than changes in temperature at least over the next twenty years. The impacts will depend on the increase in atmospheric CO2 and the reaction of trees to this increase. The potential water regulation functions of forests therefore, are receiving more and more attention in the forests and climate change discussions. The role of forests in this case is vital not only as a provider of an ecosystem service (in particular in relation to the quality of water for consumption, for irrigation, for industrial use), but also for the survival of the forests themselves. Many myths regarding the function of forests in the water cycle exist. Much more research is needed and this research should include studies on the influence of individual trees and forest stands on the local water cycle and the water cycle of watersheds.

As a rule of thumb, the original natural vegetation is the best way to regulate the water cycle. Planting trees in grassland swamps, such as one of the water catchment areas of Bogotá, Colombia, may result in rapid depletion of the water resources. On the other hand, leaving trees in dry areas may improve infiltration, soil structure and water holding capacity of the soils. Planting fast growing exotics in such areas may result in depletion of the soils. Dry areas, however, may also be affected by salination and in that case, care should be taken not to increase underground water levels, since this may increase the release of minerals in the upper soil layers.

Different plants have different water use efficiencies and these may change when exposed to more sunlight. Oak in Italy appears to be less sensitive to light conditions in its water use efficiency than a local Beech species. This may imply that beech stands may suffer less from reduced rainfall, since individual trees will react positively to lower stand densities while oak stands, on the other hand, may lose volume. In terms of effect on the regulation of the water cycle, oak may be easier to manage, since thinning of the oak stand will reduce water use more than in beech stands. Couralet *et al.*, (2010) found similar results for species in understory trees in Africa. Equal treatment of a forest stand in terms of modifications to the microclimate and water availability will have varying effects on the different species.

This relationship, however, needs more research. Knowledge of this correlation is useful for adaptation, species selection for local conditions and the estimation of the value of a species and its management system for mitigation. In Argentina and Chile, Nothofagus pumilio is suitable for mitigation due to its plasticity, while the regeneration capacity of N. macrocarpa was highly affected by the moisture gradient within canopy openings. Dietz *et al.*, (2006) in a study in Indonesia, found that tall trees may increase water evapotranspiration and decrease throughfall. This suggests that smaller trees would be better conservers of water. Of course, this is dependent on how well these smaller trees allow the throughfall and stem flow water to infiltrate in the soils, and the water retention capacity of the soils.

Apart from the effects on plant growth of individual species, climate change is likely to affect water availability at a watershed level. There has been an increase in the implementation of projects in upper watersheds with many positive benefits. However, the meta-analysis of Locatelli and Vignola (2009) clearly indicate that this is not always the case. Two strategies can be followed in such cases. They demonstrated the necessity to study the effect of trees on an experimental basis, before promoting large scale regeneration or plantations. In addition, 20 monitoring of water flow at strategic points within the watershed, taking into account current land use practices will allow for changes in the planning of species composition. In most cases, at the landscape level, control of land use and forest regeneration goes beyond the control of individuals and will need some form of stakeholder collaboration to achieve the desired results. A good case of stakeholder collaboration can be seen in the Dominican Republic, where farmers with land suitable for irrigation allowed owners of land on the ridges to occupy part of their land in return for reforestation on the ridges. It is estimated that for each hectare of irrigated land, thirty hectares are being regenerated, contributing to the availability of water for irrigation.

4

Joint Forest Management and Poverty Alleviation

Over the years, more than half of India's 76.53 million ha of forests have become degraded resulting in ecological crisis and immense hardships for the forest-dependent people in and around the forest areas. The dependence on forests is so much that over two-thirds of the rural population and half the urban population use fuelwood for cooking purposes. About a quarter of India's livestock population is almost totally dependent on forest lands. Nearly 70 percent of India's population uses traditional medicine which comes from the forests. Forest based activities are often an important source of cash income for the poor especially during lean season (MoEF 2002). The JFM Programme in India has given a new thrust and is of great relevance for developing nations who have predominantly agrarian economy and the population is dependent upon forests for subsistence.

PRIORITIES AND PARADIGM

That forests and poverty alleviation have a direct relationship is proven beyond doubt. But it is also true that poverty is higher in vicinity of higher forest cover. Among the priorities that have become prominent in development policy planning worldwide since 1980, the followings have particularly influenced the JFM policy:

- The re-conceptualization of governance: responsibility for development planning and implementation is increasingly seen as a neglected set of partnerships among state, civil society and private sector partners.
- Participatory approaches to development based on empowerment through organizational development among poor and marginalized people.
- Faith in the value of micro-enterprise based on local initiative in small production units coupled with a desire to bring these under some degree of formal state support and control (Sundar *et al.* 2001).

IMPACTS OF JFM

The JFM Programme has led to several positive impacts like:

- Improvement in the condition of forests - This is corroborated by the fact that in the past few years the overall forest cover of the country has increased by 3896 km^2 and dense cover by 10 098 km^2. Incidence of illicit felling has declined. A study carried out by the Andhra Pradesh Forest Department has indicated that between 1996 and 1999, dense and open forest covers have increased by 18 percent and 22 percent respectively. One of the more immediately visible ecological effects of JFM has been the recovery of fodder resources in JFM areas. In the study by Indian Institute of Forest Management (IIFM) of village forest committees in the Jhabua Division of Madhya Pradesh, it has been found that the average saving of a household by augmentation of fodder from the area has been Rs. 3000 per annum. The prolific growth of understorey vegetation in many instances, has led to increased bio-diversity and relatively rapid increase in wild herbivore populations.
- Increase in income - The committees have benefited from the employment generated under JFM Projects through micro planning, sale of non-timber forest products and bamboo yield, etc. Many VFCs have sustained the level of community funds, which are used for local developmental activities and personal loans, thus lessening the bondage of money lenders. In Jhabua alone the village fund is over Rs10 million as per UNICEF study. At the end of 2000-2001, total community funds under JFM were Rs.557 million (US$11.6 million) in seven states of Andhra Pradesh, Chattisgarh, Manipur, Tamil Nadu, Tripura, Uttar Pradesh and Uttaranchal.
- Reduction in encroachment - In many places JFM has helped reduce area under illegal encroachments. For instance in Andhra Pradesh nearly 12 percent of the encroached forest land (38 158 ha) has been vacated since the JFM programme was initiated. Many VFCs in Madhya Pradesh have got no encroachment resolution passed and previous encroachments were vacated.
- Involvement of NGOs - The JFM programme has led to a considerable involvement of NGOs and community-based organizations though the degree of involvement and number vary from state to state. In the six states, as per figures available from Andhra Pradesh, Manipur, Tamil Nadu, Tripura, Uttar Pradesh and Uttaranchal, 1061 NGOs are actively participating in JFM Programme.
- Change in attitude and relationship - One of the most significant impacts of JFM programme has been the change in the attitude of local committees and forest officials towards each other and towards

forests. It was unthinkable in pre-JFM days that Divisional Forest Officer will sit and discuss with the villagers while now even top forest management is easily accessible to villagers as department has accepted the role of facilitator. Several JFM related training programmes have been initiated.

The overall strategy of JFM is as described in Figure below.

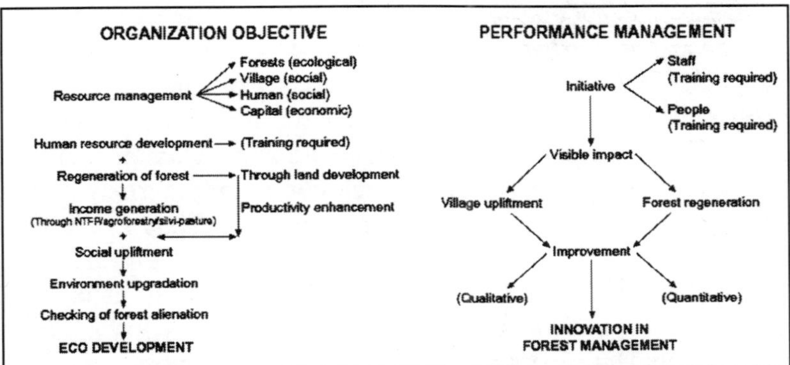

Fig. Joint Forest Management (JFM) Programme

PRESENT SCENARIO

The JFM Programme covers 14.25 million ha of forest land in 27 states through 62 890 committees. Among the many independent and isolated attempts that were made by forest departments in different states for eliciting community participation during the 1980s and 1990s, the Jhabua experiment in the State of Madhya Pradesh was unique. The Jhabua District was reeling under frequent droughts, poor productivity of natural resources - forests and agricultural lands, mass out-migration in search of livelihoods, and illegal and illicit withdrawal of forest products, when the forest department undertook the Joint Forest Management Programme. This programme not only attempted to tackle problems of forest destruction, but was also aimed at generating options for poverty eradication and employment as a crucial step for reducing pressure on natural resources.

Some states have shown a spurt of initial growth and subsequent stagnation while others who were dormant for years have shown remarkable progress in later years. Yet many others, with enormous forest wealth, despite facing extreme difficulties in managing them, have not made significant strides. In some cases, the JFM movement gathered steam with the thrust coming from externally assisted projects, but lost the momentum when the projects came to an end. The enhancement of the quality of life of forest dependent communities, through efficient, participatory, multiple-use management, equitable distribution of returns and establishing long lasting demand-spurred systems of environmental governance and justice, is possible by:

- Integrated development of land based resources along watershed approach;
- Ensuring institutional, financial and ecological sustainability;
- Establishing accountability and transparency in management practices;
- Creating institutional mechanism for empowering local communities to meet the objectives.

Forests have been identified and recognized as one of the natural resources on which the local people are dependent and extract fodder for their cattle, fuel for energy, non-timber forest produce (bamboos, cane, medicines, fruits, fibre and flosses, etc.) for their domestic needs and timber for house construction and furniture. Ecological security is the foundation of equitable and sustainable development. Forest conservation through community participation ensures ecological security and sustainable development (Roy 2003).

However, mere subsistence is not enough to ensure long term sustainability. Therefore, while ensuring sustainable forest management over a period, JFM should contribute more than subsistence needs and in the 'JFM Plus' phase it should help in raising the living standard of the community through value addition of forest produce (ICFRE 2001). The need is to have a holistic approach to forest and related natural resource management. The Dr C.H. Hanumanth Rao Committee, which looked into the working of the Drought Prone Area and Watershed Development Programme in the country during the last 30 years found that it had not worked primarily because people were not involved in the planning, implementation and management and the programmes made no provision for capacity building.

LIVELIHOOD SYSTEM

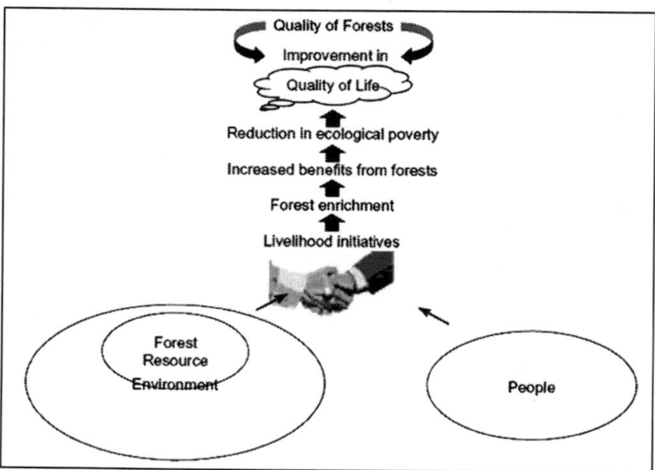

Fig. Livelihood Initiatives through Forest Enrichment (LIFE)

The 'Livelihood System' means developing the forests in a manner that the outputs from the forest provide the community with substantial economic benefit in perpetuity to attain a satisfactory level of life. The schematic representation in Figure below explains the concept of Livelihood Initiatives through Forest Enrichment (LIFE).

The following criteria for success of such initiative are envisaged based on the performance indicators:

Criteria	Parameters for success	Performance indicators
Technical	Improved ecological conditions	i. Better availability of fodder, fuelwood and other NTFPs, etc. ii. Status of regeneration by Regeneration Survey iii. Increased availability of fodder, fuelwood and timber from non-forest land
Institutional	Effectiveness of training and operationalising of principles of shared responsibility	i. Behavioural change in Forest Department staff ii. Relationship between Forest Department and community
Institutional	Performance of villages and District level initiatives	i. In terms of meeting, activities, participation, and responsibilities
Sustainability	Cost-effective eco-development activities	i. Employment generation and diversification ii. Value addition of goods
Equity	Empowerment of marginalized groups and women	i. Economic empowerment ii. Social empowerment
Participatory	Voluntary contribution	i. Increased development fund
Participatory	Better forest management	i. Rotational grazing ii. Rotational patrolling iii. Decrease in forest offences iv. Control of forest fires v. Removal of encroachment from forest land
Financial	Expected returns on investment	i. Intermediate returns ii. Expected final return

It has been found in the programme that, until an analysis of the socio-economic conditions and forest dependency for various goods and services is done, it is not possible for the Forest Department to involve local communities. Therefore a sound knowledge of their culture, tradition, social and economic conditions is a must. To take these measures, an exercise of priority setting by communities is essential. The priorities should be clear and their solution should come from the stakeholders mainly.

THE RIGHT APPROACH

The JFM Programme has to be managed in future with the following objectives:

- Suitable silvicultural practices especially the raising of nurseries like use of mulches, clonal propagation, etc.;
- Quick growing species to be developed;

- Suitable model of multi-tier plantation to be evolved and demonstrated;
- The high yielding varieties of NTFPs to be propagated and demonstrated;
- Multiple uses, high yielding varieties of fuelwood, fodder, and fruit bearing species, are to be planted for improving the economic condition of the people.

Once the JFM Programme provides the local people with alternative solutions as per the above objectives, the local communities' response to the programme will be positive and they will be convinced not to overexploit the available forest resources in the vicinity of the villages. The keenness to work in collaboration with the Forest Department shall also be more and all activities that are detrimental to forests will be reduced.

The technological inputs should stress on strengthening the existing forests by enrichment planting, soil and moisture conservation works, multiple shoot cutting in coppice species for clean boles and availability of fuelwood and fodder production linked to increased milk production.

The stress should be on the principle of care and share through socio-silviculture planning process.

The heart of enabling environment for sustainable livelihood option is participation by the forest based communities living close to forest land. It is not sufficient that forest community is 'involved'. In planning process promoted through JFM Committee, forest communities should be motivated and organized into an institution which should undertake micro-planning of public and private lands for afforestation. The planning process has to be democratic and well informed. The PRA exercises separately with men and women groups and then jointly with all are extremely useful and should cover:

- Forest resources in the village: public and private lands;
- Its status in terms of degradation;
- Uses and availability of various tree species in the forest land, other public lands and farmlands;
- Gaps between local availability of produce and the requirements of the forest community;
- Uses and availability of various grasses and shrubs;
- Uses and availability of various non-timber forest produce;
- Seasonality of employment opportunities and needs in the village.

For participatory management to be genuine, there should be enough scope for local ingenuity and innovations to contribute to management practices that would be owned by the JFM Committee. Capacity building by way of practical and theoretical training, and dissemination of success stories in vernacular language have to be the inbuilt components in all programmes (Katwal and Singh 2002).

HOW TO SUSTAIN JFM GAINS?

The approach of JFM has created a positive environment to improve livelihood systems of forest dependent communities and address the root cause of rural poverty in remote areas. Forestry alone, however, cannot provide the necessary linkages and mechanisms affecting the multifaceted problem of rural poverty. Therefore, a multi-pronged approach is needed to develop sustainable livelihood support systems under JFM.

The key areas which need attention are:

- Income generation activities for landless and women.
- Food security through Grain Banks especially in the forest fringe villages which suffer from crop failure due to low irrigation potential and difficult terrain.
- Grazing and livestock management and promotion of non-conventional energy.

NGOs can have a major supportive role in initiating and operationalising Income Generating Activities (IGAs) on collective basis. They can work as catalyst in solving marketing issues, value addition and raw material availability.

FUTURE PROSPECTS AND CONCLUSIONS

Several questions arise concerning implementation of JFM on a sustained basis. Studies have to be undertaken to find out whether the impoverished women and men, who are compelled to resort to unsustainable forest use for survival are able to switch to sustainable resource use through JFM programme.

There is a growing trend to increase the number of JFM Committees. The increase would need to be guided by the capacity of the field level forest staff and their ability to effectively coordinate and monitor the progress. Otherwise these committees remain only on paper with not much of attitudinal change. The other issues which need attention are resolving intra committee and boundary conflicts. The entire JFM mechanism has to be process-based rather than individual-based.

Given the broader concept of poverty and the broader framework for understanding poverty, as well as the global context for poverty alleviation, the framework of actions to attack poverty should be built on three pillars, empowerment, security and opportunity. These are well engraved in the JFM mandate.

Poverty alleviation is linked not only to hunger satiation but for fulfilling the basic human needs of providing shelter, clothing, clean water, education and health care. It is a paradox that where more forests exist, there is higher degree of poverty and to mitigate this problem several policy initiatives have been tried but still much is required to be done. To achieve higher progress, land has to be made productive and cattle have to be healthy. A combination of short and medium rotations and coppice system needs to be evolved for a variety of tree species being planted under JFM for faster economic gains. How productivity in the forest can be increased to meet the challenge of growing fuelwood, fodder and timber demand is the biggest challenge. The second challenge is to provide gainful employment and generate produce which can be harvested sustainably. The right approach is to manage JFM Programme by setting national objectives which should include multi-tier plantation, NTFP propagation and technological inputs which are low cost and locally adaptive. The emphasis has to be on participatory process through community based economic development, sustainable management of local resources and policy feedback.

JFM has to be taken as a means of forest development and not an end in itself. There is no short cut. Each area has specific requirements and no two JFM Committee areas can be similar in problems and solutions. Though basic approach may be the same, still implementation techniques differ. But such an integrated approach shall benefit the people, government and civil society in general. Agriculture, water harvesting and land conservation practices have to be meticulously adapted to local conditions.

Participatory forestry programmes must develop mechanisms to distribute benefits down to individuals, households and targeted groups within committees to play a meaningful role in poverty alleviation. In addition we have to put governance in 'mission mode' for a more pragmatic approach to JFM.

Every technology should have an ideology behind it to give it direction and meaning and every technology to be socially relevant should have a strategy to give it realism and experience. For achieving this in renewed perspective, the theme of any technology package under JFM programme should be to give maximum production in the shortest period; conserve forests and utilization of all that grows in the forests to meet people's needs.

FORESTRY FOR POVERTY ALLEVIATION - ROLE OF RESEARCH INSTITUTIONS

India is one of the oldest civilizations with a rich cultural heritage, but inhabited by many poor people. The country supports approximately 16 percent of the world's population with only 2.5 percent of the world's geographic area. With the turn of the new millennium, we have already crossed the one billion mark with an average density of 324 persons per km^2 . With a decadal growth rate of 21 percent, the population is projected to reach 1.25 billion by the year 2010. It is estimated that about 70 percent of the population and 80 percent of those below the poverty line live in rural areas. This includes people who live in resource-poor regions, lack productive assets, skills or capacities and those who are inadequately organised.

Since independence, India has made substantial progress in terms of improvement in basic social indicators such as health, nutrition and education. While the life expectancy has doubled, infant mortality has been halved and literacy rate has risen, a considerable proportion of population still lives in conditions of abject poverty.

Poverty, in general, is characterized by the lack of access to productive assets, basic rights and services such as health and education, besides access to information and knowledge of natural resources. Poverty thus results in exclusion and marginalisation of the people from the development process. Due to the large human and cattle population and widespread rural poverty, the natural resources of the country are subjected to enormous pressures. A major proportion of these people are directly dependent upon these resources for their survival needs.

The burden of poverty is more obvious on the rural women, already subordinated by the social structure. These women carry the burden of meeting the basic subsistence needs of food, fuel, fodder and water in the face of widening demand and supply gap, increasing environmental degradation and diminishing access to natural resources.

With the increasing recognition of the importance of forests for environmental health, energy and employment, the National Forest Policy of 1988 lays emphasis on scientific forestry research and adequate strengthening of the research base for rural and tribal development. The broad priority areas identified in the policy include improvement of productivity, effective conservation and management of existing resources and development of substitutes to replace wood and wood products. The policy also gives due consideration to the symbiotic relationship between forests and people, by emphasizing special attention to integrated development programmes. The Science and Technology Policy 2003 also identifies the need to provide food and health security for all on a sustainable basis as an important objective to be achieved through technological developments.

The research institutes therefore have a major role to play in view of the diverse research requirements for the development of "Science and Technology" in the country. While there is a strong need for conservation of our natural resources, there is an equally important need to harness their potential on sustainable basis for the benefit of the society.

The gradual realization about the widening gap in demand and supply of important forest products, inadequacy of the existing resources, their unabated degradation and the importance of conserving the complex ecosystems guided the forestry research in the country as well as at the Forest Research Institute, Dehradun. With better understanding of the changes in the state of the environment and forestry resources, coupled with the enhanced knowledge levels, the institute took up the challenges of research with much broader objectives during the last three decades. Accordingly, a number of technologies have been developed which have direct relevance to the programmes for poverty alleviation, empowerment of the masses and integrated development of villages.

IMPROVED PLANTING STOCK FOR HIGHER PRODUCTIVITY

Forest Plantations are a powerful tool in the continuing efforts of foresters to increase productivity. Increasing demand for forest products and services on one hand and decreasing land area available for forestry on the other, has necessitated raising of plantations under various combinations on farm lands and other non-forest lands.

The availability of quality planting material for such plantations through research is required.

A combination of intensive site preparation with the use of uniform, well-grown genetically-improved nursery stock, planted at uniform spacing, increases growth and yield, reduces rotation length, facilitates tending and harvesting operations and improves the wood quality. This will not only assure better economic returns to the farmers, but also reduce pressure on the remaining natural forests.

Seed Technology

The suitability and quality of the seeds have a major effect on the success of plantations raised from them. It costs almost the same to establish a plantation from poor seed, as it does from seed of high genetic potential. However, differences in the quality of plants produced and economic returns can be vast. The seed technology developed at the institute aims at production of quality seeds though various improved technological practices like seed collection, processing storage and pre-sowing treatments for effective germination. Of the various low cost and user-friendly technologies developed, a few are listed below:

- Seed processing including extraction and drying for *Azadirachta indica*
- Seed storage techniques for prolonged viability for *Azadirachta indica*, *Casuarina equisetifolia*, *Albizzia lebbek* and *Acacia nilotica*.
- Processing for improved germination in *Acacia nilotica, Albizzia lebbek, Bambusa arundinacea, Strychnos nux-vomica, Tamarindus indica* and *Tectona grandis.*

These technologies aim to realize the economic benefits by not only reducing the cost of nursery operations significantly by limiting the area of nursery to raise a calculated and desired number of seedlings, but also by production of uniform stock, thus increasing the efficiency of transplanting operations and reducing the cost. The expenditure on seed collection, extraction and processing is also reduced with optimum storage conditions.

Tissue Culture of Bamboo

Tissue culture protocols have been developed for large-scale rapid multiplication of *Dendrocalamus strictus, D. membranaceus, D. asper, Bambusa vulgaris* and *B. arundinacea.* The technology is very useful, where conventional methods of multiplication are either not available or are inadequate to fulfill the demand. This involves the use of plant tissue culture where a small plant part is cultured on artificial medium with the combination of growth regulators. A complete plant with root and shoot system is developed on synthetic medium by providing suitable light and temperature conditions.

Agroforestry Models

Agroforestry, the land use system that incorporates woody perennials with agricultural crops, help the farmers to cope with loss of crops due to drought, reduce soil and water loss, utilize off-season precipitation and meet the requirements of fodder, fuelwood, fiber, timber and other forest products, besides improved food production. Accordingly, a number of agroforestry models have been developed with different species of trees, agricultural crops and herb species, for example:

- Poplar-Sugarcane-Turmeric Block Plantation Model with benefit-cost (B/C) ratio of 3.06.
- Poplar-Sugarcane-Wheat-Chari-Potato-Maize-Bajra Block Plantation Model (B/C ratio 2.58)
- Poplar-Sugarcane-Wheat-Chari Block Plantation Model (B/C ratio 3.47)
- Poplar-Sugarcane-Potato-Barseem-Chari Block Plantation Model (B/C ratio 3.01)

SUBSTITUTES FOR WOOD AND OTHER PRODUCTS

Katha from *Uncaria Gambier*

The production of katha from the heartwood of khair (*Acacia catechu*) tree has been known for a very long time. However, shortage of khair wood prompted the institute to screen other suitable sources for making katha. One such source is *Uncaria*, a small genus of woody, climbing shrubs found mostly in tropical Southeast Asia. *U. gambier* produces the well-known gambier or pale catechu, but it has not been cultivated in India so far. The cost of production of gambier katha is much less than that from *A. catechu*. The technology would save khair trees and thus help in environmental conservation.

Jigat substitute

Machilus macrantha (Lauraceae) and *Litsea chinensis* (Lauraceae) trees are important to the survival of the agarbathi (incense stick) industry in India, which is dependent on the bark of these trees. Powder of the bark, known as 'Jigat', functions as an adhesive or binder in agarbathi. Over the years, the expansion of agarbathi industry has inflated the demand for Jigat, leading to indiscriminate felling of these trees, which is a valuable component of the evergreen and semi-evergreen forests of the Western Ghats and the north eastern states. Substitute for Jigat has been developed from agro-based biopolymers. The technology not only avoids the use of forest-based raw material but is also economically very competitive.

Natural dyes from Forest Waste

Processes have been developed for the extraction of natural dyes from some abundantly occurring plant materials of forest origin. Methods have also been developed to use these dyes on silk, wool and cotton. These dyes can be used by handloom as well garment designing industries, which export their products to developed countries like Germany and Denmark, where the use of azo dyes have been banned.

Due to environmental awareness, the natural dyes obtained from plants and animals are the dyes of 21st century. The forest biomass can be used for the production of dyes on cottage scale, generating employment for the people through value addition to the non-wood forest product and creating an additional source of revenue.

CONSERVATION AND REHABILITATION OF NATURAL RESOURCES

Rehabilitation of Mined Areas and overburden Spoils

Surface mining operations drastically affect the productivity of the land as an appreciable thickness of overburden is required to be removed to reach the ore resource.

Starting from removal of vegetation and topsoil, the ecology, socio-economic conditions and hydrology of the areas are also adversely altered. Conventional afforestation practices to revegetate mine derelict lands do not effectively rejuvenate the disrupted ecological functions, emphasizing the need for site-specific eco-restoration technologies. FRI has developed the ecorestoration technologies for surface mined phosphate mines, which have already been transferred to many companies. The salient features of the technology include an ecosystem approach towards restoration and use of ecologically and socio-economically viable species.

Cultivation Techniques of NWFP Species

The institute endeavours to undertake *in-situ* and *ex-situ* conservation of medicinal plants. Cultivation techniques of a number of economically important and endangered medicinal and aromatic plant species have been developed and transferred to various institutions, NGOs and pharmaceutical industries. The species include *Abelmoschus moschatus, C. citrates, Cymbopogon martini, Catharanthus roseus, Mentha arvensis, M. spicata, Ocimum kilimandscharicum, Rauwolfia serpentina,* and *Withania somnifera.* The technology will help to reduce pressure on forests besides being an excellent income generating activity.

OTHER LOW COST TECHNOLOGIES

Pencil Making with Hand Tools

Pencil manufacture is a complex process undertaken in modern factories, with almost all operations being carried out by mechanical appliances. FRI has developed a set of hand tools for making pencils on a cottage industry scale, with the main objective of providing additional source of income to the rural people.

The industry can be organised in community development blocks on a cooperative basis. The technology has an added advantage that it can be easily and effectively integrated with literacy programmes at the village level.

Portable Essential oil Distillation Unit

Essential oil bearing plants are very valuable as they are the sources of perfumes, cosmetics, flavouring agents and aromatic chemicals, which are also used as antiseptics, deodorants, repellents and medicines. A simple portable distillation unit of 50 kg capacity has been developed for distilling oil from essential oil bearing grasses like *Cympopogon martini*and *C. citratus* besides other leaves, roots etc. The cost of the unit is about Rs. 14 000 and is more efficient as the yield of oil is 30 percent more than the traditional ones. The distillation unit can be easily transported to the felling site/field.

Colouring and Ammonia Fumigation of Wood

There is often consumer resistance in the use of plain looking secondary plantation grown woods like poplar for furniture, in comparison to traditionally used darker decorative grained woods like teak, sissoo, rosewood and walnut. The present methods of staining and artificial grain development, based on Aniline base dyes, are not only costly, but also hazardous for health, inconsistent and develop unnatural looking grains. The process of ammonia fumigation developed by FRI gives permanent shisham, teak and walnut appearance in otherwise dull and plain looking timbers. The process is simple, inexpensive, and effective and can be adopted by small entrepreneurs, as it works out to be nearly 50 percent cheaper over conventional methods of staining.

Wood Plasticisation and Bending

Wood bending is an ancient craft and is of key importance in many industries, especially in manufacture of furniture and sports goods. The traditional steam bending technique has several limitations in quality and number of species of wood that can be bent. Recent work carried out at FRI has helped to overcome these limitations by using vapour phase ammonia plasticisation technique, enabling a wider choice of species for production of bentwood components for a variety of commercial products. The technique would economize the use of wood without affecting the functional requirements of the products, as the current practice to obtain bent wood components is from wider sections, where there is lot of wastage of timber.

Preservative Treatment of Secondary Species

Eucalypts have been planted in many states of the country to meet the growing demand for wood and has emerged as an important species for manufacture of doors/windows and joinery. It is, however, prone to termite attack, requires protection for giving long service life and is also refractory to treatment.

ACA treatment technology has been developed for the treatment of such refractory wood species to make it suitable for joinery purpose. The method has been used on a commercial scale for the treatment of eucalyptus wood for door/ window panels for which no other method was available. By this technology the eucalyptus could be economically used with the treatment cost of about Rs. 900 per m^3.

Conversion Technique for Eucalypts and Poplar

Plantation grown woods like eucalypts and poplar, though extensively available, pose problems in producing standard quality sawn and seasoned material. Major problem in its utilization is the warping in sawn timber that occurs on the saw itself and further warping and splitting in portions near the pith that occurs in subsequent air or kiln seasoning. Processing technology has been developed for their economic utilization for doors, windows, furniture and many other value added products. Recent improvements in sawing and seasoning of eucalypts and poplar have enabled them to be commercially adopted for furniture, door and windows in states like Punjab, Haryana and Uttar Pradesh with cost advantage of about 35-45 percent relative to traditional products from species like sisoo and teak, etc.

Besides these, a number of other technologies have been developed, which include utilization of juvenile wood of eucalypts and poplar for furniture, utilization of poplar for doors/windows, afforestation techniques for stress sites and agroforestry models for different agro climate regions.

ROLE OF RESEARCH INSTITUTIONS

An important constraint to the operation of the forestry research has been the lack of a method, based on the systematic application of the tested technologies for the benefit of the target group (NFRP 2000). The intended beneficiaries have not adopted technologies to the desired extent. One of the obvious reasons being inadequate linkages between research institutions and user groups, probably due to insufficient outreach efforts. Inadequate linkages, on one hand lead to lack of information on constraints in adopting technologies and the modifications needed, while on the other hand lead to insufficient impetus to augment suitable research efforts.

Table. Impact of forestry research

Direct Impact		Indirect Impact
Human Welfare Conservation and Productivity	**Institutional**	**Scientific**
• Adoption of new technologies. • Sustainable production. • Sustainable management of natural resources. • Resource conservation. • Improve environment • Extension of developed technologies.	• Capacity building • Training and education. • Networking and collaborative research. • Improved research processes. • Improved institutional effectiveness. • Improved capacity of work force. • Improved quality of knowledge adopted.	• Advances in science and technology. • Increased knowledge and understanding of the problems. • New and improved research tools and approaches. • Adaptive and applied research. • Contribution to better environment. • Contribution to better life.

Policy Interventions

The experiences over the last few decades indicate that economic growth, and targeted interventions alone are not sufficient to eradicate poverty. The essential precondition to growth should, therefore be participatory planning. The role of the target group should not only be as a beneficiary but also to act as partner in guiding the process of research and development. This requires institutional strengthening at grassroots level and bottom-up micro-planning in identifying village priorities through participatory approach. This can turn the poor communities from mere beneficiaries to active partners in the research and development process.

The National Science and Technology Policy 2003 directly addresses the problem of poverty, identifying one of its objectives as "To mount a direct and sustained effort on the alleviation of poverty, enhancing livelihood security, removal of hunger and malnutrition, reduction of drudgery and regional imbalances, both rural and urban, and generation of employment by using scientific and technological capabilities along with our traditional knowledge pool".

However, there is lack of direct attention of this issue in National Forest Policy of 1988, that addresses the issue through increased productivity, sustainable utilization of resources, and effective conservation and management of resources besides adequate strengthening of research support.

This, therefore, calls for adequate policy revision. Generation of new and adoptable technologies and screening of available technologies for their direct impact on poverty is needed. This has to be followed by widespread dissemination through networking and support for the vast unorganized sectors of our economy. The forestry research institutions thus have their role cut out for them "to ensure food, agricultural, nutritional, environmental, water, health and energy security of the people on a sustainable basis".

Stakeholder Driven Approach and Empowerment

Traditionally, the organization of research has always been highly compartmentalized. There is a strong need to rejuvenate the linkages between the researchers and the stakeholders. The lack of adequate linkages between researchers and stakeholders result in insufficient interaction and understanding of the needs. The absence of networking and sharing also results in lack of development of proper extension methods for successful transfer of technologies. The stakeholders are thus unable to benefit from the research programmes. The research institutes, therefore, need to develop adequately strong linkages with the stakeholders to understand their needs, and develop technologies. A paradigm shift in the attitude of forestry researchers towards stakeholders is needed. An integrated and holistic approach is needed with the concerted effort of all the stakeholders (Sharma 2003).

Empowerment through participation involves recognition that local people, through their knowledge and experiences, decide what is best for their development. True participation occurs when decisions by the government, services provided by the state, control of external productive resources and priority setting are carried out in conjunction with beneficiaries of these actions (Guevara 1999). James Gustave Speth in his address to UNDP in 1993 stated, "Sustainable human development is participatory. It can only be achieved when people have an opportunity to participate in the events and processes that shape their lives; where entrepreneurs, women, non-governmental organizations, and others in civil society are empowered to take initiative and participate in both open markets and effective governments, and where pluralism prevails and human rights and access of information to all parties are guaranteed". Thus the new paradigm of forestry for sustainable development must include participation, equity and environmental conservation.

Livelihood Support, Health and Nutrition

There is a strong need for collective and individual endeavours at the level of target groups as well as research institutions and development agencies to undertake activities for income generation, development and operation of community support infrastructures and creation of assets.

Forests are still called the foster mother of agriculture. In most parts of the country, forests provide an essential supplement to the nutritional status of the family. In times of food scarcity, the tubers and rhizomes provide sustenance to fight hunger. In other times, the food from forest provides essential nutritional supplements in the form of vitamins and minerals. It is, therefore, important that nutritional aspects should constitute an essential component of the package of activities at the village level.

While a number of technologies have been developed to be implemented at the grassroots level, there is need to build the capacities of hitherto marginalized groups, especially women, to become actors in development productive employment and education need to become the entry points. Research institutions can actively associate themselves with NGOs as well as government organizations to initiate projects for integrated social development, for example, community based initiatives by women's groups to start collective farming, processing of forest produce and pisciculture. The education and health interventions will converge with regeneration of livelihoods and productive employment to ensure positive impacts on the society especially the marginalized sectors.

Increased availability of even a single resource, say, for example water or fuel would lessen the women's burden and provide them with time and opportunity to take up income generation activities such as gum collection, crafts and fodder production. Capacity building of women and youths can also be done through developing cadres of para-professionals, not only for primary health care but also for activities like rainwater harvesting, developing and maintaining biogas units, etc. Research institutions have a very important role to play in this field by organizing training workshops to train these para-professionals.

Reviving and Strengthening Traditional Knowledge Systems

While issues of sustainable agriculture, health, nutrition and education can be addressed through strengthening of community based organizations, there is a strong need to focus on improving, adapting and reviving local technologies not only for health and nutrition but also for water harvesting in traditional ponds, use of traditional methods of agriculture etc. For example, while the modern high yielding varieties require high inputs in-terms of fertilizers, pesticides and water requirements, the traditional pest and drought resistant varieties can provide equally good harvest with low inorganic inputs. R&D institutions need to actively associate themselves with programmes for revival of such practices, besides developing strong linkages with community and NGOs. Research institution can also pay a lead role in setting up seed banks of important species and in adapting traditional technologies of rain-fed agriculture to rehabilitate degraded lands.

The importance of traditional knowledge in the field of health care needs no explanation. The importance of documentation of this knowledge is being talked about, the world over. Building on traditional knowledge in the field of medicinal plants, will not only help to preserve it for posterity, but also aim at optimal use of forest resources besides helping the community to take control of the health of its members. Research Institutions can help to take up research on traditional systems of healthcare so as to contribute to fundamental advances in health care, help to develop effective commercial products as well as appropriate norms for their standardization and validation. They also can have a key role to play in development of technologies for value addition to indigenous resources for their optimal utilization. This will have a direct impact on the health care as well as sustenance at the grassroots level.

The process of globalization is leading to a situation where the collective knowledge of the societies, normally used for common good is converted to proprietary knowledge for commercial profit of a few. Research organizations need to actively associate themselves with development of IPR systems to protect scientific discoveries and technological innovations arising out of tradition and indigenous knowledge.

Education, Empowerment and Capacity Building

Education not only refers to formal education or literacy status, but also enhancement of the knowledge level of the society as well as the individuals to become partners in development. The research institutions can build upon the emerging concept of para-professionals to enhance capacity of the community to access facilities and serve the poor. Para professionals are people from with in the community or group who can be trained through capacity building to deliver the services to the society as well as provide feed back to the institutions. For example in accessing technology, to serve poor peoples needs, and exposure to and understanding of that technology, its appropriateness, implications and sustainability is important. The R&D institutions thus have a role to play not only in development of appropriate user-friendly technologies but creating enabling environment and building leadership within community.

Genuine forestry development manifests itself in the alleviation of poverty and the basic problems could be addressed by a four-pronged strategy, which could be outlined as:

- To ensure education, training and capacity building of stakeholders.
- To augment research and its validation.
- To develop extension and technical cooperation programmes for the validated research results.
- To ensure appropriate participation and empowerment through suitable methodologies and policies.

Ultimately the research output can only be judged by its impact on development. The research institutions must therefore reorient themselves to "bridge the knowledge gap, i.e., disparity in the capacity to generate, acquire, disseminate and use scientific and technical knowledge, which is the most vital difference between the rich and the poor" (IDRC 1991).

CONCLUSIONS

The creation and implementation of plans for sustainable management of forests through integrated forest strategies is one of the central mechanisms for poverty alleviation programmes. It is not only important to view forest resources as effective potential tool for poverty alleviation but also to understand the connection between poverty and deforestation. Poverty is often viewed as one among the many causes of degradation of natural resources. With no better alternatives, forest destruction and degradation become short-term solutions to many of the burdens imposed by poverty. It is equally true, however, that for poor people who live in and near forests, and those who live in places where woodlands have been destroyed or degraded, silvicultural interventions can be effective in regenerating and rehabilitating such forests. This could be instrumental in reducing poverty by providing increased and sustained employment in the production and processing of timber and non-timber forest products. We must now agree that unless poverty is alleviated, the forests will continue to be lost. It is therefore, important that the natural resources especially forests are recognized as primary poverty fighting assets. An integrated and holistic approach is important to amalgamate the ideas from different disciplines and sectors to achieve successful sustainable forest development.

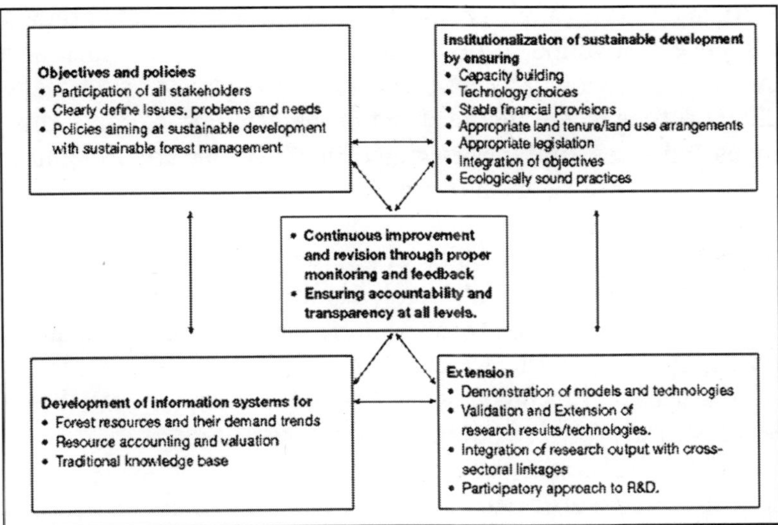

Fig. Integrated forest strategy

Forestry research institutions can play a leading role in the development of integrated forest strategy. As one of the major stakeholders, the research institutions can take up the challenges to develop ecologically sound practices, provide training for capacity building, ensure validation and extension of technologies, as well as develop information systems to integrate research outputs with cross-sectoral linkages.

Creating an integrated strategy requires the interactions of forests with other sectors of the economy. The forest research institutions have the knowledge, experience and expertise to take up this challenge. Such institutions can thus play a pivotal role in utilizing the tool of knowledge to attain economic power by playing a decisive and beneficial role in improving the well being of all sections of our society. They can have a central role in raising the quality of life of the people, particularly the disadvantaged sections of the society, in creating wealth for all by utilizing natural resources in a sustainable manner and by protecting our environment.

FOREST FOR POVERTY ALLEVIATION: CHHATTISGARH EXPERIENCE

The ever increasing anthropogenic pressure and archaic managing institution on forests, particularly in tropical countries, have led to depletion of the vegetation, land degradation, distortion of hydrological cycle and consequent decrease in productivity resulting in poverty and misery. Problem becomes acute in forest fringe areas where there is neither enough landfit for cultivation nor industries to provide employment. Due to disadvantaged geographic location (DGL) syndrome, Human Development Index (HDI) is at its lowest in these areas. People still practise primitive subsistence agriculture with very low productivity and to meet their growing food grain requirements they opted for more extensive cultivation including shifting cultivation or encroachment in forest areas. Degradation of forests as a result of exploitation for fuelwood and illicit felling of trees is yet another facet of the same problem. In this process steep slopes and areas unfit for growing of annual crops are brought under the plough, creating near ecological disaster.

The water-balance in the situation described above is totally upset by the destruction of vegetation. Distortion of the hydrological cycle and consequent decrease in productivity per unit area leads to enhanced poverty and misery. Poverty and illiteracy coupled with malnutrition are again the main causes of increased population growth, which further accentuate the natural resource degradation process. Thus poverty in these areas becomes both the cause and the effect of natural resource degradation (Sharma 1999).

It is a debatable point whether poverty with its pressure to survive or affluence with its pressure to consume leads to environmental degradation but it is more than obvious that poor people cannot be signatory to conservation if it is in conflict with their survival needs. The 21st century challenge is to facilitate a devolution of greater authority to forest-based communities while minimizing conflicts, and to support new partnership among communities, government and the private sector to ensure the meeting of community needs, forest resource conservation and sustainable use. Clarifying forest use rights and responsibilities and creating adaptive policies and programmes that allow for intensified access controls can lead to more sustainable forest management. This requires appropriate institutional arrangements (Poffenberger 1996).

The fundamental role of forests in national development, poverty alleviation and food security has again been recognized at the World Food Summit (WFS 1996) and (2002), the United Nations Millennium Summit (2000) and the World Summit on Sustainable Development (WSSD 2002), where in following priorities for action have been identified:

- Developing the contribution of forests and trees to poverty alleviation and food security by (i) securing subsistence needs; (ii) generating income and provision of forest inputs and services to non-forest income generating activities; (iii) raising the bargaining power of the poorest people through better access to natural capital assets; and (iv) reducing vulnerability to environmental and economic shocks.
- Enhancing the provision of essential public goods by forest and trees by recognizing the multiple benefits of forests to protect watersheds, mitigate climate change and act as reservoirs of biological diversity.

For the Asia-Pacific region poverty is a serious concern. The twin problem of rural poverty and continuing degradation of forest are posing a real challenge. Therefore in this chapter the role and potential of forest sector in reducing poverty will be examined in terms of identifying the opportunities and capturing and strengthening them by revisiting the fragile link and vexed issues between the ecological security and livelihood security of the dependent people. During this journey we will try to learn the ABC of forestry (where "A" stands for appropriate entitlement regime, "B" for benefit sharing arrangement and "C" for conservation) because our strong belief is that to reach "C", we have to move through "A" and "B" sequentially. Finally, a proactive and people friendly framework with actual operational modalities and lessons learned from the newly created state of Chhattisgarh, India, will be cited which is a modest attempt to reconcile the dichotomy of threat perceptions arising out of conservation-development orthodoxy by taking into account the human sensitivities in terms of the felt needs of the people, their social norms, beliefs and systems borne out of history, culture and traditions.

SETTING THE SCENE

Poverty is a complex phenomenon. Apart from the macro economic problems, the degradation and restricted access of the poor, to the available material and environmental assets also fundamentally trap the poor in their circumstances. Poverty in turn leads to further degradation of the natural assets and circumscribes the limited access the poor have over the natural resources. This constitutes the vicious circle of poverty particularly in the resource rich but underdeveloped areas with undefined and inequitable access of the poor to the common property resources.

Agenda 21 of the Earth Summit (1992), reiterated that to provide all persons urgently with the opportunity to earn a sustainable livelihood there should be an integrating factor that allows policies to address issues of development of sustainable resource management and poverty eradication simultaneously.

Livelihood Security

A livelihood comprises the capabilities, assets (stores, resources, claims and access) and activities required for a means of living: a livelihood is sustainable which can cope with and recover from stress and shocks, maintain or enhance its capabilities and assets, and provide sustainable livelihood opportunities for the next generation; and which contributes nett benefits to other livelihoods at the local and global levels and in the long and short term. Accordingly, sustainable livelihood approach permeates the entire concept of People's Protected Area.

As a corollary to this, to begin with, food security, health cover and dependable wage labour become areas of prime concern. Goods and services from the forest ecosystem more importantly non wood forest products (NWFPs) originating from diverse sources ranging from large plants to micro flora consisting of heterogeneous products, constitute a critical life line for poor forest dwellers by providing family sustenance and livelihood. Due to their recurrent availability on annual/seasonal basis and immense socio-cultural, economic, environmental and industrial development potentials, NWFPs hold a promise for developing interesting mechanism for sustainable livelihood.

Food Security

Food security envisages adequacy, stability as well as economic and physical access to food to all people at all times. There may be enough food but if the poor don't have access to it, the food security will not be complete.

Besides ameliorating the soil and water conservation regime, which adds substantially to the enhanced agricultural food production, forest products reinforce the food security in many ways. Human history corroborates that forests have always been a source for large number of non-wood forest produces which directly contribute to the food basket of the people in form of edible fruits, flowers, gums leaves, roots, tubers etc. During lean agricultural season, even the agricultural communities supplement their food requirement from the forests. Furthermore, income generated from sale of surplus NWFPs enables the poor to have access to food.

In the Indian rural context where mixed farming is in vogue, cattle are important component of the socio-economic set up. In the complex chain of food web, cattle are secondary food producers too. They derive their food from the forests, which they convert into animal proteins in form of milk, fat, meat and other dairy products to be used by human beings. Although excessive grazing pressure, more importantly the one beyond the carrying capacity, has been viewed as a constraint by foresters, PPA endeavours to accommodate multiple use of natural resource on sustainable basis so that food resources are available to the people as well as cattle.

Health Cover

Forest have been source of invaluable medicinal plants since the time man realized the preventive and curative properties of plants and started using them for human health cover. Even when no synthetic medicines existed, our forefathers had been depending on herbs and medicinal plants and their derivatives to cure common ailments. Our age-old traditional Indian System of Medicine (ISM), one of the most ancient medicine practices known to the world, derives maximum of its formulations from plants and plant extracts that exist in the forests.

The general forest degradation process adversely affects the resources base of medicinal and herbal plants both in terms of quantity as well as quality. Rural poor, whose dependence on these products is very heavy, are the worst sufferer. The problem is compounded by market demand driven harvesting without any concern for regeneration and conservation. In this process essential regenerative component of a plant like bark, roots flowers and fruits are indiscriminately collected leading to degradation and depletion and even demise of particular species, if proper remedial measures are not taken. Many important medicinal plants like *Rauvolfia serpentina*, *Curcuma caesia, Dioscorea* spp., *Chlorophytum* spp., etc. are becoming rare and some of them are critically endangered. It is estimated that 10 percent of all plant species and 21 percent of mammal species are currently endangered in India.

Augmenting Rural Employment

Land and water are two most important natural endowments but they are finite. However, with judicious mix of interventions like development of irrigation facilities, application of improved and modern agricultural practices and creation of other income generation activities based on non-destructive use of locally available natural resources, the possibilities of creating dependable wage labour can be enhanced. The initial entry point activity would be to create awareness among the local people about their latent strength, availability of natural resources and potential of using them on sustainable basis by technological up gradation. Broadly, this may be achieved by developing the hardware of the system comprising of physical activities eg. water harvesting structures, drinking water facilities, common facility centers, village level processing units, storage units etc. On the other hand, interventions like raising awareness, bridging credibility gap, confidence building, soliciting meaningful participation of the local communities and empowering them, creating income generation opportunities through skill development etc. would constitute the software of the system.

Keeping in view that for the poor communities, economics precedes ecology and conservation and in tune with their priority, there has to be sharp focus on productivity enhancement so that they can reap better economic returns. For those who possess land, a programme is initiated to upgrade the productivity of the agriculture system by land shaping, constructing stop-dams, tube wells and making available to them other inputs for better crop, husbandry practices, so that the land under rain fed condition is transformed into double cropped or in certain cases even triple cropped land. Enhancing the productivity of the land provides good harvest as well as gainful employment round the year. All these on-farm activities can be adopted as a package for improving the socio-economic conditions of the people who otherwise practise subsistence agriculture. For landless persons off-farm activities leading to income generation can be devised by using forest biomass and other resources. In the first category, activities like NWFP collection, rope making, honey collection, nursery etc can be considered, whereas in the latter category schemes like, mushroom cultivation, shop keeping, grocery, etc. can be envisaged (Sharma 1997).

NWFPs, with their attendant instrumentalities, play a meaningful role in bringing sustainability to the system because the employment generation from these enterprises is around 20 million man days per year, which is approximately half of the forestry sector. NWFP related activities take care of both the unemployed as well as underemployed and NWFP based small scale enterprises can further strengthen the linkage of the socio-economic base on account of:

- Low capital and low energy requirements.
- Proper utilization of local renewable resource and technological know-how.
- Checking migration from rural to urban areas; and
- Being a family activity it provides satisfaction of "creation".

These are some of the illustrative and by no means exhaustive interventions, which can provide a conducive environment for socially acceptable and dependable wage labour.

To translate this premise, the concept of People's Protected Area (PPA) has emerged which by targeting on broad range of goods and services in terms of physical, material, human, social and environmental assets in conjunction with appropriate entitlement regime, PPA envisions a proactive and people's friendly framework so that it becomes people's pool of assets for meaningful poverty alleviation and their enhanced well being alongwith conservation.

A NEW STATE IS BORN

As per provisions of the Madhya Pradesh Reorganization Act, 2000, a new state of Chhattisgarh was born on 1st November 2000. The 16 districts of erstwhile Madhya Pradesh spread over an area of 135 000 km^2 constituted the new state. The state has rich endowment of natural resources in terms of minerals, forests and water bodies. Important mineral deposits are of iron, coal, bauxite, uranium and diamond. The state has a forest cover of around 44 percent, which represents diverse tropical flora and fauna. Mahanadi, Shivnath, Son, Arpa, Kharoun, Hasdeo and Indrawati are the main rivers. With an average annual rainfall of 125 cm. the state produces some of the best rice varieties and hence aptly christened as "Rice Bowl" of the country.

With the population of slightly over 20 million people, Chhattisgarh has got a high proportion of scheduled castes (SCs) and scheduled tribes (STs).

Whereas for the country as a whole the SCs and STs population is 23.6 percent, in Chhattisgarh the combined population of SCs and STs is 44.7 percent, consisting of 32.4 percent of STs and 12.3 percent of SCs.

The state has around 20 000 villages of which 9500 villages are forests adjacent, which have more than half of the population belonging to tribal groups. *Gonds* form the largest proportion (55 percent) of the tribal population. The tribal and other backward classes are concentrated in the hilly southern and northern districts where the hillocks are covered with forests. Most of the SCs are located in the central and north central parts of the state, particularly in the districts of Raigarh, Kawardha, and Mahasamunda.

Chhattisgarh has got 59 772 km^2 of forests, which have been classified as sal, teak and miscellaneous forests including bamboo forests. Some of the best sal forests of the country are in this state. Apart from timber, these forests provide many non-wood forest produce (NWFP) like tendu leaves, sal seeds, mahua flower and seeds, amla, harra, gum, lac, tamarind and mahul leaves, etc. Besides these, several important medicinal plants are also found here. These non-wood forest produces are important source of income and also serve as food supplement during famine and scarcity, which are quite recurrent in this area. More than 50 percent of the people living in and around forest area depend for their subsistence on forests. Furthermore, forests find a place in the rich socio-bio-cultural matrix of the local populace. It is against this background that Chhattisgarh has enunciated a new state forest policy (CGSFP) 2001, with the following main objectives:

- Unlocking of the vast array of forest resources on sustainable basis for enhanced well being of local people by converting these open access resources (OAR) into community controlled, prioritized, protected and managed resources.
- A shift in accent from major to minor forest produces, from crown to multi tier forestry and from flagship species to smaller denizens of the forests.

MAHANADI PROTECTED AREA (MNPA)

To reconcile the conservation-development orthodoxy, the concept of People's Protected Area (PPA) has emerged which by targeting on a broad range of goods and services in terms of physical, material, human, social and environment assets in conjunction with appropriate entitlement regime, envisions a proactive and people's friendly framework so that it becomes people's pool of assets for meaningful poverty alleviation and their enhanced well being alongwith conservation.

The People's Protected Areas delineate sites containing predominantly unmodified natural systems, under management to ensure the long term protection and maintenance of biological diversity, whilst providing at the same time a sustainable flow of natural products and services to meet local community needs. While JFM has been perceived as a forest department programme in which people participate, PPA involves a paradigm shift from forest management to integrated eco-system management in which socio-economic well being is the goal and forests are viewed as a means to achieve it. Moreover JFM is process oriented and does not lend itself to becoming a target and product oriented programme (Saxena 1997).

Furthermore PPA makes liberal use of social capital, which includes norms of trust, reciprocity and network that facilitate mutually beneficialy cooperation in community. The affirmative social capital acts as a catalyst to promote sustainable forest management as well as socio-economic development of the people (Anderson 1998).

To translate above premise, 32 PPAs each extending over 15 000 to 20 000 ha covering more than 300 villages, have been established as a model of "conservation through" use. Thus every forest division in the State has one People's Protected Area. It is proposed to extend the coverage gradually to other forests of the State.

Goal of MNPA

The area identified under MNPA is inhabited mostly by poor forest dwellers including tribals and other Primitive Tribe Groups (PTGs). In spite of more than 50 years of planned economic development in the country, positive aspects of the same have not percolated down to these areas. The human development index (HDI) in these areas is therefore very low. Under this scenario the primary goal of the MNPA is:

- Creating a hunger free zone.
- Preventing distress migration from this area; and
- Ensuring job opportunities to every capable person.

Further, the MNPA will focus the highest level of conservation efforts for protecting the pure and wild germplasm of the wild buffalo, the "state animal", besides tiger. MNPA will also focus on conservation of natural features, habitat/ species management, i.e conservation through active management, protected landscapes and sustainable use of natural ecosystems.

Learning from the experience of Asian economic miracle of rapid growth with reduced inequality, mechanism will be developed for transmitting the gains throughout the economy and particularly to the poor.

Focusing on the role of water as a precious and finite resource, which has a direct relationship with the food security, livelihoods and human health, MNPA endeavours to understand the relationship between forests and freshwater and to manage the forest for sustaining the productivity of uplands without affecting the soil and water on which they depend. It has become all the more important because more than 3 billion people on earth do not have access to clean water.

Of the more than 3 million deaths that are attributed to polluted water and poor sanitation annually, more than 2 million are children in developing countries.

Area and Location Map of MNPA

The total area of MNPA is 516 000 ha encompassing 572 villages falling in Dhamtari and Raipur districts of Chhattisgarh. The details of the forest area and number of villages in division/range is given in the following table:

Division	Range	Forest village	Rev.e village	Total village	Forest area in km^2				Tot. revenue area in km^2	Tot. area in km^2
					R.F.	P.F.	Orange area	Tot. forest area		
Dhamtari	N. Singpur	6	53	59	233.67	1.86	0	235.53	846.46	
	Keregaon	6	42	48	203.7	20.64	0	224.34		
	Dhamtari	3	67	70	78.01	44.9	0	122.91		
	S.Singpur	5	18	23	245.25	1.82	0	247.07		
	Dugli	7	0	7	207.79	0	0	207.79		
	Birgudi	17	82	99	192.9	0	0	192.9		
	Nagri	10	21	31	181.25	0	0	181.25		
	Sankra	4	23	27	155.2	0	0	155.2		
	Sitanadi (S)	10	0	10	261.57	0	0	261.57		
	Risgaon (S)	19	8	27	296.98	0	0	296.98		
E. Raipur	Gariaband(Part)	0	35	35	0	158.25	4.16	162.41	575.85	
	Nawagarh	1	22	23	0	236.79	10.22	247.01		
	Dhawalpur	2	24	26	0	256.96	0	256.96		
Udanti	Mainpur	0	31	31	0	218.08	14.6	232.68		
	Kulhadighat	2	7	9	0	231.37	0	231.37		
	Tauranga	5	29	34	32.16	209.97	0	242.13		
	Udanti (S)	1	12	13	109.94	137.65	0	247.59		
TOTAL		**98**	**474**	**572**	**2198.42**	**1518.29**	**28.98**	**3745.69**	**1422.31**	**5168**

VISION, VALUES AND MISSION

Vision

Network of People's Protected Areas (PPAs) as poor people's pool of assets for sustainable livelihood by unlocking forests for people through integrated ecosystem approach.

Values

- Highest respect and concern for people and their traditional knowledge.
- Care and share.
- Capacity building at all levels.
- Up gradation of local technologies including use of information technology.

Mission

- Community based participatory mapping and management plan.
- Appropriate resource assessment methodologies.

- *In-situ/ex-situ* conservation and propagation.
- Forests and freshwater.
- Non-destructive harvesting.
- Grading, processing, value addition, certification, eco labeling and marketing.
- Bio-cultural diversity conservation, biodiversity prospecting and bio-partnership.
- Ecotourism.
- Carbon Sequestration.
- Entrepreneurship development.
- Public-Private Partnership (PPP).
- Gender sensitivity.
- Equitable benefit sharing arrangements (BSA).
- Improved food security and health cover.
- Enhancement of social capital.
- Monitoring evaluation and people's indicator of SFM.
- Enabling policy and legal framework.

A multidisciplinary team, headed by a senior forest officer has been constituted and it is expected that the first preliminary report will be available by the end of this year, the contents of which will be firmed up in a seminar/ workshop of all concerned stakeholders and subject matter specialists.

FROM PROMISES TO PERFORMANCE

Creating Enabling Environment

The National Forest Policy 1988, the State Forest Policy 2001, the JFM Resolutions and directives as well as decisions of the Supreme Court of India provide the basic policy framework of PPA.

The legal framework for the People's Protected Areas flows from the Indian Forest Act 1927 and the 73rd Amendment of the Indian Constitution, viz, Provisions of the Panchayats (Extension to the Scheduled Areas) Act 1996, which *inter alia* provides for conferring the endowment of ownership rights of NWFPs on Panchayati Raj Institutions (village level institutions). In line with this, the state is endowing the ownership in consonance with the following principles:

- Harvesting of NWFP will be on non-destructive basis.
- The members of the *Gram Sabhas* will be free to collect NWFP for their own consumption.
- The Manner, frequency and intensity of NWFP collection for any use other than *bona-fide* domestic use by the members of the *Gram Sabhas* will be in accordance with the prescription of a management plan prepared by *Zila Panchayat* in conformity with the guidelines as may be notified from time to time.

Besides the remunerative wages paid to the NWFP collectors, the nett profit from the trade will be shared among the stakeholders on an equitable basis. Under the existing system of tendu patta (*Diospyros melanoxylon*) trade, the nett income generated by the collection and trade of tendu leaves is distributed in following proportion:

- 70 percent to the primary collectors.
- 15 percent for the development of NWFP and regeneration of forests.
- Balance 15 percent for infrastructure development/Cash Payment (CGFD 2003 a).

This modality aims at developing positive stakes of all concerned so that rather than working at cross-purposes, each one is motivated to contribute towards a common goal. The state has further developed an equitable benefit sharing mechanism for timber and bamboo also as evident from the salient features of various JFM resolutions given below:

- Every family of the JFM committees will be entitled to receive *Nistar* forest produces subject to their availability.
- All the forest committees shall be eligible to get 100 percent of the forest produces obtained from time to time, from mechanical thinning and cleaning of rehabilitated area and cleaning of bamboo clumps in degraded forests as per prescriptions of micro-plan/working plan, on payment of expenditure incurred on harvesting.
- Forest products equivalent to 15 percent of the amount calculated (of timber/bamboo) by deducting the expenditure incurred on harvesting (of timber/bamboo) from the total value of forest product or cash equivalent to that shall be given to the Forest Protection Committee (FPC).
- Forest produces equivalent to 30 percent of the amount calculated by deducting the expenditure incurred on harvesting (of timber/ bamboo) from the total value of timber/bamboo obtained on final felling in plantation/rehabilitation of degraded forests, shall be given to the Village Forest Committee (VFC).
- Forest produce equivalent to 100 percent of the amount calculated by deducting expenditure incurred on borrowing capital (both cash or kind) and expenditure incurred on harvesting from the total value of timber/ bamboo obtained on final felling in plantation of degraded forests shall be given to the Village Forest Committee (VFC). The entire operation will be as per prescription of micro plan, prepared on location specific basis.

As a result of these benefit sharing arrangements (BSA), the forest dwellers have received around Rs. 600 million as their share form timber, bamboo and tendu leaves, besides wage employment opportunities to the tune of Rs. 1200 million. Establishment of "herbal state"

- Chhattisgarh has been declared as "Herbal State".
- CG MFP Federation has prepared a comprehensive project for sustainable development, conservation and utilization of medicinal plants with active participation of locals including traditional *vaidyas*.
- A State Medicinal Plant Board has been constituted to formulate policy for medicinal plant resource conservation and sustainable utilization.
- An interdisciplinary task force has been constituted to prepare and implement the project.
- Herbal dispensaries are being established in interior forest areas.
- FRLHT (Foundation for Revitalization of Local Health Traditions), Bangalore, Shristi herbals, Raipur and other NGOs have been associated.
- Srishti Herbal University has been opened in the private sector.
- State Biodiversity Action Programme has been formulated.

Chhattisgarh State Minor Forest Produce (T&D) Co-op. Federation Limited

Chhattisgarh State Minor Forest Produce (Trading and Development) Cooperative Federation Limited (CGMFP Federation), Raipur, is an apex organization of approximately 2 million forest produce gatherers comprising 913 primary cooperative societies and 32 district unions. It is also the nodal agency for all aspects relating to management, development and trade of minor forest produce/non-wood forest produce sector in the state.

Under the three-tier cooperative structure the primary cooperative societies have been constituted with the membership of actual pluckers and chairman of the society is chosen from amongst the members only. At present there are over 10 000 collection centers spread over the length and breadth of the state and the annual turnover of the trade is over Rs. 2500 million. In line with the time tested philosophy of care and share and to ensure that harvesting of NWFP has essentially to be on a non-destructive basis, serious and concerted attempts have been made to convert these poor people from gatherers to owners.

With proper research focus, sustainable harvest and appropriate utilization pattern including godowning, processing and marketing, the turnover from NWFP may cross even the Rs. 10 000 million mark a substantial sum by any standard for the rural poor.

The state has initiated appropriate measures through the CGMFP Federation for sustainable utilization and long term conservation of all NWFP found within the forests of the State. Some of the measures taken are:

- The rate for pruning (branch-cutting) was raised from Rs. 10 per standard bag to Rs. 20 per standard bag in order to improve the quality of tendu leaves as well as the earning capacity of the tribals and other economically backward communities.

- An amount of Rs. 292.7 million was distributed as bonus to the tendu patta collectors for the year 1998, (50 percent of the profit of Rs. 585.4 million).
- 30 percent of the profit i.e. Rs. 175.6 million was invested in the development of infrastructure in the rural areas and rest 20 percent of the profit i.e. Rs. 117.1 million was ploughed back to the development of forests.
- For the year 2001, out of the total profit of Rs. 445.15 million, Rs. 311.6 million (70 percent) was given back to the tendu patta collectors as bonus, Rs. 66.7 million (15 percent) was invested in the development of infrastructure development in the rural areas and Rs. 66.7 million (15 percent) was used for the development of the forests. This year about Rs. 450 million will be paid as incentive bonus, out of the profit of the tendu patta trade.
- 474 225 quintal of sal seed was collected in the year 2001 and an amount of Rs. 152 million was distributed to NWFP collectors. This year around Rs. 200 million will be disbursed to the sal seed collectors.
- State Government has delegated the powers of Registrar, Cooperative Societies, to the Executive Director of CGMFP Federations and Deputy Registrar to the Managing Director/DFO (Dist. Union), to create an enabling environment for the smooth conduct of business by the primary cooperative societies and the district union.

These interventions by the MFP Federation have led to increased assured wages to the NWFP gatherers in the interior areas where there are no employment opportunities otherwise. In this process, they got around Rs. 930 million in wages in the year 2001 alone. Series of such well-orchestrated interventions can substantially enhance the well being of the poor forest dwellers.

Integrated Ecosystem Approach

In order to formulate a people friendly framework for poverty alleviation, sustainable forest development and biodiversity conservation through integrated ecosystem approach, two models have been evolved namely 'Dhamtari Model' and 'Marwahi Model'.

Dhamtari Model

Outcome of execution of a programme is determined not just by new policies but by institutions as well. Presently in the forest fringe the institutional framework responsible for developmental administration consisting of various line departments is not truly effective because there are no infrastructure facilities in the interior areas and programme implementers look for softer options. Under these constraints, forest fringe areas suffer from disadvantaged geographical location (DGL). That is why the poverty map of the country approximately tallies with the forest map. It is in this context that an innovative development administration system should be thought of. Forest management, on account of their physical presence in those areas could be considered as practical alternate agency for facilitating, coordinating and in some cases even executing programmes of the line departments responsible for the socio-economic development.

This agency can be designated as Forest Fringe Area Development Authority (FFDA).

The rural development funds for any administrative unit can rationally be divided between the existing agency and FFDA management as per the following formula.

If X is the number of the villages within the forest or 5 km of the forest (fringe area); Y is the number of villages outside the fringe area, A being the total funds available and B and C, the funds to be spent by the FFDA and DRDA respectively, then

$B/A = X/(X+Y)$, and $A = B + C$

i.e. the funds made available are in direct proportion to the ratio of the villages existing in the fringe area to the total number of villages. This apportionment must be laid down at the State headquarter level to obviate any chances of favouritism, with the *Panchayat Raj* Institutions (PRIs) having the option of allocating more funds than this share to the fringe areas.

In Chhattisgarh about 50 percent of the villages with population of more than 10 million people are located within 5 km from the boundary of the forests. In a large number of these villages, poverty is rampant due to variety of reasons. The State Government is committed for overall development of these poverty stricken people and conservation of their natural resource base.

Forest department has been made a nodal agency for integrated development of all the 401 villages situated within 5 km periphery of forests in Dhamtari District, the first venture of its own kind in our country. Thus, villages situated in the fringe areas of the forests have been brought under the umbrella of forest administration for implementing the integrated ecosystem approach at landscape level by convergence of all development schemes. The State Government has also made budgetary provision for these developmental activities.

Marwahi Model

It has been observed especially in case of donor driven forestry projects, that during time span of the project implementation wage employment increases but immediately after the project is over, the employment falls down, adversely affecting the sustainability and credibility of the programme. This can be taken care of by evolving a carefully crafted withdrawal strategy. If the project can create durable assets, even after the project is over, the assets so generated can continue to provide regular job employment/income. This basic principle has been operationalised in the Marwahi Model.

In line with this, for rehabilitation of degraded forests of Marwahi forest division, an innovative scheme has been conceptualized. This scheme is unique in the sense that it gives enough flexibility to carry out works at many appropriate sites, which will activate many forest committees.

In the working plan total degraded forests of the division has been kept in 19 treatment series. In these series, no coupes have been demarcated. Accordingly, as per provisions of working plan the rehabilitation work is to be done by the committees after taking up 30 hectares of treatment unit at one place after preparation of micro plans.

In working plan, about 60 000 ha area has been identified as the degraded forests. This year 5000 ha of degraded forest has been taken up for treatment. Considering 60 percent as workable area, 3000 ha area is being worked this year and around 100 sites have been selected for the treatment. For area wise preparation of micro plans experts and NGOs as well as Samitis are being involved. After careful scrutiny and screening, 40 NGOs, forest experts and research scholars have been selected and the job of preparation micro plans is in progress.

ECOTOURISM

Ecotourism is basically low impact utilization of the forest ecosystem services, which has a great potential for socio-economic development in remote areas.

Besides being a great repository of biological diversity, forests of Chhattisgarh contain various sites of archeological, cultural and religious importance. Some of these unique combinations of natural and cultural heritage can form nuclei of ecotourism Accordingly three circuits of eco tourism namely Raipur, Barnawapara, Turturia, Sirpur (Raipur District), Achanakmar (Bilaspur District) and Kanger valley (Jagdalpur District) have been started which are controlled and managed by the committee of forest dwellers. All the benefits accrued are passed over to the communities

APPROPRIATE RESOURCE ASSESSMENT METHODOLOGY

At present, there is no systematic and reliable data regarding the availability of non-wood forest products. Local people, on the other hand have sound knowledge about the different kinds of NWFP species and their natural occurrence in the adjoining forest areas.

The State aims at assessing and preparing a data bank on quantity, quality and value of various NWFPs existing in an area, by active involvement of local people, using local resources and local technology. Using the sample survey methods and based on sound statistical principles, a Comprehensive Community Based Participatory Mapping and appropriate Resource Assessment

Methodology (RAM) has been developed. The resource assessment methodology adopted includes laying of stratified systematic sampling plots, regeneration plot for knowing the regeneration status of the area and medicinal plot for getting an accurate idea about production potential and regeneration status of medicinal plants.

For estimation of the production potential and regeneration number of medicinal and NWFP plants of each species in northeast quadrant of the sample plot will be counted.

Thereafter, the weight of useful part of five plants (in the case of bigger species such as *baibidang, malkangni, marorfalli, safed musli, kali musli,* etc.) and 50 plants (in the case of smaller species like *kalmegh, bhui aanwla, bhringraj, punarnava, dudhi,* etc.) will be taken in wet and dry conditions. If by any chance the number of plants in this quadrant is less than the above numbers, the available quantity will be used.

All these observations will be entered into prescribed proformas specially designed for this purpose. After collecting the various data and details mentioned above, different results like number of trees/ha, species wise volume/ha regeneration status, species-wise number and yield/ha. of medicinal plants and NWFP species etc. will be obtained. For the computation of growing stock prevailing conversion factors will be used. Details and data collected about medicinal plants and NWFP species will be used in deciding about the target species and their treatment prescriptions etc.

SUSTAINABLE FOREST MANAGEMENT (SFM), FOREST CERTIFICATION AND NONWOOD FOREST PRODUCE (NWFP)

Global awareness towards sustainable forest management and economic implications along with market forces and consumer preferences for quality forestry products is necessitating a management paradigm that ensures the sustainability of forests. Since last decade, many international/national processes have been labouring hard to formulate Criteria and Indicators (C&I) of Sustainable Forest Management (SFM). However, in spite of extensive scientific, social, economic and political debate, no consensus has been arrived at as to what constitutes the SFM. This is further compounded because, varied objectives, value systems, temporal and spatial scales coupled with inherently long time-period to determine the efficacy of methodologies used for defining and ascertaining sustainable forest management defy clear and acceptable formulations.

On account of difficult conceptual issues, it is evident that sustainable forest management with all its attendant paraphernalia for formulating the criteria and indicators will remain an impracticable proposition, at least in developing countries, where poverty is rampant. Therefore, while maintaining health and vitality of the forest ecosystem, what is needed is to evolve a package of proactive and people's friendly minimal damage forest management practices which could contribute incrementally towards sustainable forest management or avoid those practices, that are clearly destructive and simultaneously enhance the well being of people. This indeed will be a practical approach towards attaining the desired goal rather than perusing the elaborate and illusive theoretical matrix of SFM. Unless it is so, all laudable initiatives of JFM or SFM will wilt before bloom.

Currently a wide range of actions are underway concerning certification. Although main emphasis to date has been on timber and timber products, attention has recently expanded to include pulp and paper products. There is an urgent need of developing certification system for non-wood forest products too.

The forests of Chhattisgarh state are being managed as per prescriptions of the Working Plans which are based on sound principles of forest management and have got the approval of the Government of India. Working plan provisions have been made mandatory for the working of the entire forest area, as per the directions of Supreme Court of India.

There are a total 6687 Forest Protection Committees/Village Forest Committees in the state, which are managing approximately 28 890 km^2 of forest area of the state. The collection of NWFP is being done on the 'principles of ecological sustainability, economic sustainability and social sustainability'. There has not been tradition of using chemical fertilizers or insecticides in the forest areas of Chhattisgarh. Thus the NWFP of forest areas of Chhattisgarh are organically grown.

In recognition of the fact that certification of NWFPs can provide remunerative price to the collectors/ growers and side-by-side the industry will get quality raw material in conjunction with sustainable forest management, a one-day Workshop on "Certification of Non-wood Forest Produce including Medicinal, Aromatic and Dye Plants" was organised at Raipur on 9 April 2003. After general deliberations, four specific working groups were constituted to look into the entire gamut of issues such as the fair average qualities (FAQs), quality assurance through laboratory testing facilities, forest management issues pertaining to certification, and basic points of certification process in forest management (general principles).

Towards this end, we are constituting an autonomous certifying agency consisting of representatives of following stakeholders:

- Growers/primary collectors
- Village level committees
- University/research organizations
- State Government
- Autonomous bodies
- Expert bodies like C.I.M.A.P., C.D.R.I., or any other National institute/ body, having expertise in medicinal plants/NWFP.

PUBLIC PRIVATE PARTNERSHIP (PPP)

Chhattisgarh Forest Department has initiated many schemes to ensure active participation of local people, NGOs and industrial houses for the sustainable livelihood of the poor through integrated ecosystem approach including reclamation of degraded forests. The economic implications of tangible goods of the forest ecosystem are proposed to be enhanced through high quality inputs in form of cash as well as kind, non destructive harvesting, value addition, processing and marketing and equitable benefit sharing, in partnership with the dependent people, entrepreneurs and other interested organizations.

After formation of the state the following steps have been taken up in this direction:

RECLAMATION OF DEGRADED FORESTS

Keeping in view the fact that about 80 percent of the people living in the vicinity of the forests, in one way or the other are dependent on the forests, the State is trying its best to arrest the pace of degradation and simultaneously rejuvenate the depleting forests for providing sustainable livelihood. However looking to the extent of degraded forests and existing budget allocation, this is proving to be a Herculean task.

The rason d'etre for exploring other avenue is: undue long time frame to rehabilitate these areas by the Government efforts alone with limited financial resources, the quality and general performance of the previous efforts being what they are and the lessons learned from modalities of the private sector participation in various countries including China. Hence it is felt that the task of restoration of degraded forests can be accomplished faster with the active participation of private sector and accordingly arrangements are being worked out within the ambit of Government of India JFM Resolution December, 2002 and the guidelines issued relating to National Afforestation Programme (NAP). This model of Public Private Partnership (PPP); with symbiotic relationship among the various stakeholders can contribute substantially towards greening of degraded forests without actually giving private sector any access to forestlands. In fact this could be a win-win situation for the Government, the forests dependent people and the private sector.

Forward and Backward Linkages of NWFP and Medicinal Plants

Till now villagers were getting returns from nationalized NWFPs like *Tendu patta, Sal Seed, Harra, Gums* only. But after establishment of PPA non-nationalized NWFPs including medicinal plants are being mainstreamed for reinforcing the livelihood security of the people.

Memorandum of Understanding (MOU) incorporating buy-back-guarantee agreement between village *samities* and traders/herbal industries are being encouraged and forward and backward linkages are being established. The overall outcome is very promising and it is expected that in next NWFP season many more such MOU's will be signed.

Mining and Revegetation of Rocky Areas

As already described, Chhattisgarh is a storehouse of many important minerals, most of which are found in forests. The mining strategy in these areas should be based on "green technology" with built in safeguards for mitigating the negative impacts. Possibilities are being explored to plough back a part of revenue generated from mining for forest development and generation and also for the welfare activities of the local population.

Some parts of degraded forests in the State are rocky in nature. Such areas are completely devoid of any vegetation. There is no biological productivity derived from such areas and consequently, villagers living nearby do not derive any benefits. However, these areas can be brought under green cover by digging pits and or by opening up of the rocky area and then filling these gaps by enriched soil. This will not only revegetate the rocky areas but the forests committee can also use the stony byproducts for construction works or for road repairs etc. An initiative in this regard is being taken up in the State to create livelihood sustainability for the people as well as to bring such rocky patches under vegetal cover. Furthermore, such interventions will improve the hydrological cycle of the locality.

Forest development agency (FDA)

The scheme entitled National Afforestation Programme (NAP) has been formulated by merger of four centrally sponsored afforestation schemes of Ministry of Environment and Forests with common funding pattern to be implemented through Forest Development Agencies.

FDAs' decentralized institutional structure would allow greater participation of the community both in planning and implementation of the appropriate afforestation programmes. This would ground the people-centered approach in afforestation programmes and provide a firm and sustainable mechanism for devolution of funds to JFMCs for afforestation and related activities.

Organic unity in this structural framework will promote efficiency, effectiveness, accountability through decentralization and devolution of authority and responsibilities, both physical and financial. Village will be reckoned as a unit of planning and implementation and all the activities under the scheme will be conceptualized at the village level. The two-tier approach apart from building capabilities at the grassroots level would also empower local people to participate in the decision making process.

- FDA will be constituted at the territorial/wildlife forest division level and shall have a general body and an executive body. FDA will be a registered society under the Societies' Registration Act.

- At the grassroots level, the Joint Forest Management Committee (JFMC) will be the implementing agency. In the proposed structure, one JFMC will cater to a village.

In Chhattisgarh, 29 FDAs have been constituted. 22 FDA proposals have been sent to Government of India out of which 12 FDA proposals have been sanctioned. Under FDA, different types of activities have been taken into account such as assisted natural regeneration, fuelwood plantation, silvipasture development, bamboo plantation, mixed plantation of trees having NWFPs including medicinal, aromatic and dye plants (MADP).

5

Green Trade and the Impact of Non-wood Forest Products in Joint Forest Management

The Indian Forest Policy of 1988 (MoEF, 1988) and the subsequent government resolution on participatory forest management (MoEF, 1990) emphasize the need for people's participation in natural forest management. The policy document asserts that local communities should be motivated to identify themselves with the development and protection of the forests from which they derive benefits.

Thus, the policy envisages a process of joint management of forests by the state governments (which have nominal responsibility) and the local people, which would share both the responsibility for managing the resource and the benefits that accrue from this management.

Under joint forest management (JFM), village communities are entrusted with the protection and management of nearby forests. The areas concerned are usually degraded or even deforested areas. However, in Andhra Pradesh and Madhya Pradesh all village fringe forests can come under JFM. The communities are required to organize forest protection committees, village forest committees, village forest conservation and development societies, etc. Each of these bodies has an executive committee that manages its day-to-day affairs. Non-wood forest products (NWFPs) have a key role in JFM efforts. With the increasing awareness of their economic potential and growing concerns for the sustainability of the resources and the distribution of the benefits derived from them, various state governments have taken over control of a number of NWFPs.

IMPORTANCE OF NWFPS IN JOINT FOREST MANAGEMENT

Non-wood forest products are important to JFM efforts for a number of reasons. First, NWFPs are integral to the lifestyle of forest-dependent communities. They fulfil basic requirements, provide gainful employment during lean periods and supplement incomes from agriculture and wage labour. Medicinal plants have an important role in rural health (Prasad and Bhatnagar,

1991). In parts of West Bengal, communities derive as much as 17 percent of their annual household income from NWFP collection and sale. According to J.Y. Campbell, small-scale forest-based enterprises, many of which rely on NWFPs, provide up to 50 percent of the income for about 25 percent of India's rural labour force.

Second, NWFPs have a decided advantage over timber in terms of the time needed to achieve significant volumes of commercially valuable production. Timber production is a long-term endeavour, and in many areas timber harvesting may not be ecologically desirable. Moreover, many NWFPs become available even in the earliest stages of rehabilitation of degraded forest areas.

Third, at the national level over 50 percent of forest revenue and about 70 percent of forest export revenue comes from NWFPs, mostly from unprocessed and raw forms. Thus NWFP management has clear ecological, social and economic benefits. Managing forests for multiple products including NWFPs and adding value to them at the local level are two of the most pressing challenges facing the JFM programme. In attempts to optimize the production of multiple products to meet the objectives of the various stakeholders, due attention should be paid to the potential for sustainable production of NWFPs in forest management efforts, including JFM arrangements. The true spirit of JFM gets translated only when forests are also managed to meet the people's needs.

THE MOVE TOWARDS STATE CONTROL OF NWFPS

Traditionally, the collection of NWFPs has been of low intensity and generally sustainable. However, as the economic potential of NWFPs has become apparent, the intensity of collection has increased and more significant infrastructures for trade and processing have developed. This has raised concerns for the sustainability of the resources and the distribution of the benefits derived from them.

In reaction to these concerns, a number of state governments have taken over the control of a number of NWFPs. The state regulations bringing certain NWFPs under monopoly trade. Some of the explicit objectives for state monopoly of NWFP trade are:

- To prevent unscrupulous intermediaries and their agents from exploiting NWFP collectors:
- To ensure fair wages to collectors:
- To enhance revenue for the state:
- To ensure quality;
- To maximize the collection of produce.

In most cases, trading is controlled through state-owned institutions such as state forest development corporations, federations, cooperatives and tribal societies. In Orissa, however, where the Forest Produce (Control and Trade) Act of 1981 provides the scope for a state monopoly on certain selected forest products, the state also has the option to give monopoly leases for collection and trade of forest products. In fact, the state has granted monopoly rights for 29 NWFP items to a private company, Utkal Forest Products Ltd. Under this agreement, the local people who collect NWFPs are required to sell their collected materials to the company's agents at preset prices that are lower than those they could have obtained by selling directly to processors. It is noteworthy that some of the 29 items yield very insignificant amounts of revenue yet have nevertheless been taken under the state monopoly.

Collection and Trade of Tendu Leaves Under State Monopoly

The first NWFP brought under state control was tendu leaves (*Diospyros melanoxylon*), used to wrap traditional cigarettes (*bidi*). This tree species is found in abundance in tropical deciduous forests, on wastelands to some extent and even on private holdings. Tendu collection was monopolized by Madhya Pradesh in 1964, followed by Maharashtra (1969), Andhra Pradesh (1971), Bihar (1973), Gujarat (1979) and Orissa (1981). The monopolization of tendu was rapidly followed by similar procedures for other economically important NWFPs, including sal seed (*Shorea robusta*), gums and myrobalan (*Terminalia chebula* and *Terminalia bellerica*).

Table. State trading regulations promulgated by state governments

State	Regulations	Implications
Andhra Pradesh	Andhra Pradesh Minor Forest Produce (Regulation of Trade) Act, 1971 Andhra Pradesh scheduled areas	Trade in NWFPs is declared state monopoly whether ownership is with government or not
Bihar	Bihar Kendu Leaves (Control of Trade) Act, 1973 Bihar Forest Produce (Regulation of Trade) Act, 1984	Bihar State Forest Development Corporation operates as state government agent for collection and marketing of kendu leaves, sal seed, mahua (*Madhuca latifolia*) and harra
Gujarat	Gujarat Minor Forest Produce (Regulation of Trade) Act, 1979	Minor forest products identified include timru leaves (tendu leaves), mahua flowers, fruits, seeds and gum
Himachal Pradesh	Himachal Pradesh Resin and Resin Produce (Regulation of Trade) Act, 1981	Resin, bamboo and *Acacia catechu* (khair) collection through Himachal Pradesh Forest Development Corporation Ltd
Madhya Pradesh	Madhya Pradesh Vanopaj (Vyapar Viniyam) Adhiniyam, 1969	Items under monopoly include tendu leaves, sal seed, harra and gums; Madhya Pradesh Minor Forest Produce (Trade and Development) Federation acts as agent of state government
Rajasthan	Rajasthan Tendu Leaves Act, 1974	Rajasthan Tribal Area Development Federation collects and markets NWFPs
Orissa	Orissa Forest Produce (Control of Trade) Act, 1981 Orissa Kendu Leaves (Control of Trade) Act, 1981	Collection and sale of NWFPs are monopolized by Forest Department and leased to Tribal Development Cooperative Society which in turn delegates to an individual; collection and trade of leaves are handled by Forest Corporation

Before Madhya Pradesh adopted a cooperative structure for tendu leaf trade in 1988, the collection of leaves as per official records ranged from 6 to 7 million standard bags. This was reduced to around 4 million standard bags per year after 1989. The reduction did not result from a lack of resources, but rather from the rejection of leaves that would previously have been collected but were not of high enough quality for the cooperatives. However, local manufacturers of *bidi* cigarettes have been known to buy additional tendu leaves directly from collectors.

In Madhya Pradesh, collectors share in profits through a bonus plan at the end of each season. In Orissa, which is also rich in NWFPs, the collectors get only wages for collection; the bulk of the profit goes to the Forest Development Corporation, which has been given monopoly rights by the state government (Agragamee, 1997). As in Madhya Pradesh, collection of tendu leaves is being limited by a desire to collect only the best produce.

Table. Phases in collection of tendu leaves in Madhya Pradesh

Phase	Total period (*years*)	Collection per year (*million standard bags*)	Growth rate
1965-1980	15	2-3	+1.0
1981-1988	8	6-7	+5.93
1989-1996	8	4	-1.87

Table. Trends in collection of some monopolized NWFPs in Orissa

Product	Collection (*tonnes*)		Increase or decrease (%)
	Before state monopoly	After state monopoly	
Tendu leaves	36 000 (1967-1973)	35200 (1979-1985)	-2.02
Sal seeds	200 000 (1977)	60000 (1987)	-70
Lac	32 000 (1961-1970)	16000 (1981-1986)	-50

Monopoly of Broomstick Grass and other NWFPs in Orissa

Broomstick grass grows wild in most hilly tracts of the Raygada district of Orissa. Tribal women, through JFM forest protection committees, have protected this grass from grazing and fire and have obtained income by selling it. However, since the trade in broomstick grass has come under state control its collection and sale have been much reduced. While the women get 1.5 to 3 rupees (US$0.03 to $0.06) per kilogram of broomstick grass, the company holding the monopoly (Utkal Forest Products Ltd) is making profits of as much as 600 percent. Reduced collection has also been observed for *T. chebula,* gums and mahua (*Madhuca latifolia*) flowers since these were taken under state monopoly. However, collection of *Buchananea lanzan* and *Chlorophytum tubersum* (safed musli) has increased, perhaps to the point of unsustainability, in response to high commercial demand.

DOES THE STATE MONOPOLY MEET ITS OBJECTIVES?

While the objectives of creating state monopolies are laudable, a comparison of the declared objectives and the actual situation suggests that on the whole the results are not favourable. Although state forest revenues have increased, the forest-dependent communities do not appear to be reaping benefits in terms of wages, socio-economic conditions or gender equity, and the cost to end users has continued to increase. Intermediaries have not been eliminated, but have been replaced by agents of the monopoly leaseholders. Moreover, forest degradation has not been halted, and destructive harvesting has been observed in some cases where the prospect of short-term profit has obscured care for long-term sustainability.

Table. Declared objectives of NWFP trade monopoly and the field situation

Stated objectives	Actual situation
Welfare of forest-dependent communities	
Ensuring access to forests (implied)	Restricted by agents and subagents of government, e.g. in Orissa. (Agragamee, 1997)
Ensuring fair wages through prompt	In Orissa, bulk profits are being apportioned by and just payments intermediaries/state government; in Madhya Pradesh, payments are not only delayed, but are inadequate because of the managed collection of predetermined quantity
Elimination of intermediaries	Intermediaries are agents and subagents of monopoly leaseholders, e.g. in Orissa (Agragamee, 1997)
Preventing exploitation of NWFP collectors	Where government alone does the marketing, it is inefficient; where marketing is left to private trade, it is exploitative (Prasad and Saxena, 1996)
Ensuring better socio-economic conditions	Perceptible improvement in economic conditions is not in sight
Maximizing collection to ensure more wages	Collection is regulated and thus the full advantage of the produce is not available to the NWFP gatherers, as illustrated by declining collection in Orissa and Madhya Pradesh
Maintenance of benefit sharing and gender equity	The National Commission on Women reported very low payments to women, partly because of ignorance and partly because of women's fear of being denied collection access to forests (Prasad and Saxena, 1996)
Sustainable forest management	
Sustainable harvesting	Agents and subagents of monopoly leaseholders are interested in enhanced return and are unmindful of the long-term impact of destructive harvesting (*Emblica officinalis, Buchanania lanzan, Diospyros melanoxylon. Madhuca* spp.)
Forest protection	Collection of sal seed, mahua (*Madhuca latifolia*), etc. is associated with extensive forest fire to clear the ground of leaf litter and to ensure clear visibility of the produce on the ground
Natural regeneration	Regeneration is very poor because of destructive harvesting, grazing and fire
Forest productivity	Forest degradation continues
Forest revenue to state	Forest revenue to state has increased tremendously
Protection of end users' interests	End users continues to pay more because the market is controlled by others

IMPACT OF THE MONOPOLY OF NWFPS ON JOINT FOREST MANAGEMENT

In many states the membership of JFM committees and primary forest produce collectors (PFPCs) societies are not the same. While membership in forest protection committees (FPCs) and village forest committees (VFCs) is open to all families of the village, PFPCs are made up of collectors only. FPCs and VFCs are entrusted with the conservation and development of forests and NWFPs, but no such responsibilities are assigned to the NWFP collectors. Thus, while the PFPCs get the bonuses and other monetary benefits accruing from the sale of NWFPs under monopoly, they are not held responsible for management of the resource.

On the other hand, those who are tasked with protection of the resource do not share in the benefits of its exploitation. The two types of organizations therefore need to be integrated so that the interests of members conserving and developing the forest resource and those of NWFP collectors do not clash. This could be done by inclusion of members of JFM committees in the PFPCs. Alternatively, part of the profit from NWFPs could also be given to JFM committee members who are not part of the PFPCs.

SUMMARY AND CONCLUSIONS

It appears that the objectives of bringing NWFPs under state monopoly - to reduce exploitation of tribal people and other forest-dependent communities and to promote sustainable management of NWFPs - are not being realized. State control over the trade of NWFPs has often resulted in compounded problems of restricted access to resources and non-remunerative returns to the collectors. NWFPs are one of the keys to successful joint forest management, but if local people who are engaged in the arduous task of collecting NWFPs are not able to get fair wages even when the trade is handled by government-appointed agencies, JFM may not be a viable tool in the achievement of sustainable forest management.

REQUISITES FOR THRIVING RURAL NON-WOOD FOREST PRODUCT ENTERPRISES

For millennia forests have provided local communities with food, medicine and fibres and with income from trade in these items. Only in the past few centuries has trade in timber and pulp overshadowed these various commodities. Enterprises based on non-wood forest products (NWFPs) have attracted attention for their potential to make forest use more sustainable, both because they extend the range of forest benefits and because gathering and processing activities can be managed by communities near the forest resource, with a greater portion of the end-product revenues returning to those who manage the resource. General discussions of the potential of NWFPs and their international markets have often overshadowed the actual experiences of many community-based enterprises and their success stories.

The main requirements for successful non-wood forest enterprises and sustainable forest management are strong local institutions and clear, supportive national policies (particularly related to land rights and marketing), rather than international markets. Experience suggests that for almost any forest product, international demand alone is more likely to yield only short-term economic gain and to lead to forest destruction.

NWFP ENTERPRISES TODAY

Although it is still a small fraction of the world trade in timber products, trade in NWFPs is far from insignificant. World trade in medicinal plants alone reaches US$10 000 million annually (Freese, 1998). Rural economies benefit from the impact of NWFP trade on national economies; for example. Indonesian exports of rattan and other NWFPs exceeded US$134 million per year, and Indian trade is estimated to be US$1 000 million (Freese, 1998).

1. There is no separate trade classification for medicinal plants of forest origin: however, the majority of medicinal plants are harvested from the forest.

For rural economies and forest management. the great significance of NWFP enterprises lies in regional, national and local markets. The numerous small enterprises involved in NWFPs can foster broader-based economic growth than large-scale timber operations. Earlier this decade in Zimbabwe, such small forest enterprises employed an estimated 237 000 people, compared with 16 000 employed in the country's formal forest industry. Because NWFP harvesting often employs women and minorities to a greater extent than large-scale timber operations, it has potentially greater equity benefits.

Table. Comparison of risks and benefits for markets at different levels

Type of market	Relative risk to producer	Nature of trade
Local rural markets	Low: low transport costs; market preferences easily accessible	Possibly slow growth; demand for forest products declines as exposure to cheap manufactured products increases
Urban and national markets	Medium: transport costs higher; less information available on market preferences	Potentially fast growth because of urban migration; for traditional products, urban markets can be larger than international ones
Regional markets (neighbouring countries)	Somewhat higher: higher transport costs; more information needed on markets, export-import regulations and tariffs	Potentially strong within regions with shared ecological and cultural features; these deserve more study
International markets	High: frequently requires intermediaries to gain information on product standards; more sophisticated market preferences and international trade rules	Historical tendency to boom-bust sequences of quickly rising demand, followed by swift decline as technologies develop for production of cheaper substitutes

Some long-standing systems manage non-wood resources by producing products that combine trade and subsistence use. For example, in the late 1800s Sumatran farmers planted local damar (*Shorea javanica*) and native fruit-trees, mainly durian (*Durio zibethinus*) and lansat (*Lansium domesticum*), in their forests. The damar's clear resin was exported for use in paint and varnishes, while the fruits were consumed locally.

NWFPs are important to industrialized as well as developing economies. Vantomme (1998) notes that in temperate regions pines have been tapped for resins for over two thousand years. Pine nuts, long used in many parts of the world, have become part of gourmet cuisines in Asia, Europe and North America. In the United States' Pacific Northwest alone, trade in non-wood forest resources (Christmas ornamentals, mushrooms and other edible products, and medicinal products) reaches at least US$200 million annually (Hansis, 1998).

FACTORS SUPPORTING SUCCESSFUL NWFP ENTERPRISES

Successful enterprises and sectors based on non-wood forest resources and products show a pattern: most involve local incentives, clear rights governing forest use, healthy local institutions and links with stable markets. Emphasis on quality products is also a factor for success (EWW, 1999). Local processing can increase returns through value added, while tapping local knowledge can enhance the production and processing as well as the marketability of products.

Local Processing

A key to success in NWFP enterprises lies in adding value to the non-wood resource through local processing, which returns a greater portion of the final price to the people who manage the resource. Local processing can preserve items, reduce postharvest losses and enable the product to reach more distant markets.

In the Philippines, local processing was essential to the success of the Kalahan Educational Foundation (KEF), a local organization that helped the Ikalahan of north central Luzon gain legal control over their ancestral forest through the Philippines' first community forest stewardship agreement (Rice, 1995), which gave communities an incentive to manage the forest sustainably. KEF provided a forum for establishing local priorities and mechanisms for achieving them. In 1980, KEF members established a food-processing centre and began testing methods for making jellies and jams from forest fruits. The goal was to capture 10 percent of the high-end market in Manila (Rice, 1995). By 1995, the processing centre was providing 150 local families with most of their cash income. A KEF inventory of local wild guava resources on 500 ha of forest reserve at current harvesting levels found few signs of negative impact. The processing effort has in fact encouraged planting of more trees on private farms. The success of this endeavour encouraged KEF to reinvest the profits from the processing centre locally; they were used to build a local health centre and a secondary school.

Improving Marketing Operations

Long and opaque market chains can be an obstacle to NWFP enterprises. However, it should not be assumed that market chains are always inefficient or exploitative; in many instances, intermediaries perform an indispensable service by absorbing short-term risks and by providing access to markets that are hard to enter. Community-based enterprises can help local producers take advantage of opportunities to improve market chain operations. In Nepal, the trade in medicinal plants has long involved rural collectors, yielding an estimated US$10 million in annual income. However, collection was haphazard and frequently illegal, and there were often huge disparities between what collectors and intermediaries received. Uncontrolled harvests and the existence of a black market for the plants had brought several species to the brink of local extinction.

Table. Ways in which forest producers can increase their incomes.

Means	Example
Increase production	
Practise thinning, weeding	Remove competing vegetation
Improve harvest techniques	Harvest more selectively Learn about better equipment use
Improve postharvest storage	Construct ventilated storeroom Dry fruits for shipping
Control pests	Monitor insect damage Use biopesticides rather than chemicals
Increase the product value	
Improve quality	Sort produce by quality grades
Process materials	Assess processing options Install processing facilities Package products
Obtain higher prices for the product	
Organize for greater strength with traders	Group sales to traders at standard prices based on quality
Employ group marketing direct to consumers	Collectively rent transport for taking products to market Obtain credit for members Store product for off-season price benefits

Table. Comparison of market intermediaries' role and producers' options in two villages in peninsular Malaysia

Feature	Village A		Village B	
Products	Coconut, cocoa		Fruits: durian, petai	
Investment to enter market	Overhead, labour		Minimal	
Product price	Relatively fixed		Flexible	
Average selling price (*US$*)	Coconut	Cocoa seed	Durian	Petai
- To intermediary	0.06	0.24/kg	0.40	5.60
- Direct to consumer	-	-	1.60	7.20
Marketing channel	Well established		Informal	
Potential for producer to enter market	Low		Good	

In the remote Jumla district, the medicinal plant jatamashi (*Nardostachys grandiflora*), which grows in alpine meadows and woodlands, is a major cash earner; its bitter, hairy rhizome yields a valuable essential oil. Several projects have helped communities to enter this industry where market access is difficult. The Humla Oil Project involved local partners in the Humla Conservation and Development Association (HCDA) with the regional Asian Network for Small-Scale Agricultural Bioresources (ANSAB) and EnterpriseWorks Worldwide (an international non-governmental organization [INGO] based in Washington, DC, formerly Appropriate Technology International).

The project recognized that community members could gain more control over the jatamashi market by processing their own product, so it established two factories for distilling the essential oil; the factories cost US$35 000 and were completed in 1996. Collectors sold the roots to Humla Oil for processing. Humla Oil worked with them to map local jatamashi resources and document regeneration of jatamashi and other commercial plants. In the Raya area, this led the community to start a rotational harvesting system. The Humla Oil factories employ ten workers and do business with 590 collectors, who in 1997 gained a total income of US$24 000. The Humla Oil Project has broadened its operations to include essential oils from other plants, including juniper berries, sugandhawal and sunpati. Its emphasis on quality has allowed its products to enter the large Indian market, and some are shipped to two firms in the United States.

Exploring regional markets. Regional markets, serving cultures that share similar product preferences, may offer the greatest return to government research efforts and investment in NWFPs and processing techniques. The following example shows how research that builds from local use can identify regional niches and the potential for sustainable ventures.

In the Kékoldi indigenous community of Costa Rica, shopkeepers traditionally sold medicinal tea and medicines made from the wood chips of the bitterwood tree (*Quassia amara*). Harvests involved cutting the trees to metre-high stumps, which could resprout for future production (Ocampo, 1994). Bitterwood attracted the interest of researchers at the Tropical Agricultural Research and Higher Education Center (CATIE), who found that it had commercial potential as a biological insecticide. The researchers found that local demand for the traditional medicine was low enough to permit additional sustainable harvests in the 118-ha area, and accordingly explored techniques for processing the wood chips to make the bio-insecticide for domestic and regional sales.

However, the researchers discovered that the amount needed to enter export markets would exceed the level of sustainable harvest in the Kékoldi reserve and concluded that supplementary sources would need to be found throughout Central America before the enterprise could proceed.

In Thailand, Wanida *et al.* (1993) explored the potential for processing and marketing catechu (*Acacia catechu*) - a tree native to India - to satisfy traditional Thai demand and also for export to India, where it is a popular chew snack. This study of processing, market and resource management implications, including local use and regional market standards, exemplifies a well-integrated approach to product research and marketing.

Alliances for green marketing. "Green marketing" targets the increasing number of consumers who are willing to pay a premium for products that are certified as sustainably produced and environmentally safe. If this premium returns to the producers of the product, it will help defray some of the costs of sustainable management of NWFPs, thereby helping to ensure their economic and environmental sustainability (Freese, 1998). Conservation International in the United States and Social Ventures Network in Europe have forged networks for green marketing, and groups such as Enterprise Works Worldwide (EWW) and the Biodiversity Conservation Network have provided technical and management support services.

Support for Local Knowledge

A project in Nepal demonstrated how NWFP enterprises can build on local knowledge and skills. Over the past decade, women's cooperatives in Nepal's eastern Makalu-Barun National Park and Conservation Area developed trade in allo cloth, a traditional product made from giant nettle (*Girardinia* spp.), a tall perennial that grows abundantly there. In 1990, the Nepali Government and the Mountain Institute provided technical support for marketing and business operations for cooperatives in four villages. A marketing study found allo cloth to be promising for sale to high-end tourist markets in Kathmandu. Its tweed look and flexibility make it suitable for a range of market uses from coats and shawls to upholstery.

The project fostered private distribution links at either end of the plane journey to Kathmandu. The distributors matched orders to supply and assumed short-term transport costs. A clear price list with acceptable profit margins set for each stage in the market chain helped to prevent exploitation of producers. To ensure continued supply of the giant nettle and minimal disruption of the wild population, the women explored techniques for allo cultivation.

The allo enterprise fostered respect for local traditions and knowledge. Traditional product design and household production featured prominently, with the recognition that high-end consumers value the environmental and cultural basis of production if it is presented effectively (Nicholson, 1995).

Policy Support

A coherent government policy can help enterprises manage NWFP resources and enter nearby markets.

Nepal's Community Forest Management programme allows communities to receive a portion of royalty payments for raw materials collected from their lands, which would otherwise go to the central government. This royalty for local management has proved to be a powerful incentive for communities in the area to take on forest management responsibilities (EWW, 1999). The Humla Oil Project described above, for example, has convinced the local district forest office to accept its plan for community-based management of the resource under this programme.

In the 1960s, farsighted planners in mountainous Himachal Pradesh, India recognized the advantages of the state's diverse native forest produce. Because of the state's marginal soils, small average farm size and poor access to inputs, field crops had limited potential. However, the planners recognized that a variety of fruits and nuts already harvested by farmers were better suited to the mountain soils. The nearby Delhi market offered good prospects for fruit sales. By promoting fruit production, the state encouraged trade, soil conservation and better nutrition among its rural households. The Directorate of Horticulture gave farmers information on nearby markets and promoted quality control from harvest to packaging, consistent product standards and appropriate infrastructure (such as gravitational ropeways to bring fruit from the orchard to the road). With these incentives, farmers shifted from field crops to more profitable forest species and supplemented wild resources by planting fruit-trees on their own land. The government thus expanded enterprise opportunities and laid a foundation for economic and environmental stability.

CONCLUSIONS

The experiences described above and others like them have provided lessons about the elements contributing to successful NWFP-based enterprises. The following five points are central.

- *Support communities through clear land tenure and policy support.* Where local groups are well organized and can control forest access, rural enterprises tend to fare better. A clear sense of group identity, cooperative behaviour and established rights to the resource can all help.
- *Start with local markets.* Local markets are easier to enter and monitor than foreign markets, which often require heavy capital investment and large product volume and which tend to be vulnerable to product substitution. Enterprises may diversify to larger markets if such diversification is feasible in terms of sustainable harvests, product quality and investment requirements.
- *Focus on quality products and building management and entrepreneurial skills.* These elements can be supported through coalitions involving local partners, local and national NGOs and international technical organizations.

- *Support NWFP enterprises through policies facilitating credit and trade.* Coherent government policies that support NWFP enterprises are needed, including mechanisms to make credit more available to small enterprises (such as the recognition of stands of commercial tree species as collateral) and the removal of counterproductive price controls. To inspire policy-makers to support rural enterprises with a coherent policy framework, FAO has proposed better accounting of the economic importance of NWFPs, including a system for grouping NWFP trade statistics within existing commodity classification systems (Chandrasekharan, 1995).
- *Make the most of local knowledge and resources.* Maintaining cultural integrity remains an underappreciated element of forest sustainability, particularly in remote communities and upland areas. Researchers in silviculture, marketing and processing should consider the best available knowledge from traditional as well as scientific sources to optimize forest management and the contribution of NWFPs to the lives of rural people.

GREEN TRADE ORGANIZATIONS: STRIVING FOR FAIR BENEFITS FROM TRADE IN NON-WOOD FOREST PRODUCTS

Green trade organizations aim to secure sustainable supply systems and to ensure benefits to local people from national, regional and international trade. These organizations promote trade practices that adhere to the principles of sustainable development in all its facets: ecological sustainability (use and conservation), economic sustainability (productivity) and social sustainability (equity).

The main handicap of those involved in production and trade of non-wood forest products (NWFPs). from indigenous peoples' associations to private-sector enterprises, is the lack of market information and market access and the difficulty of obtaining information on appropriate processing technologies for their products.

Green trade organizations can help improve access to national and/or international markets for NWFP producers, especially those in developing countries, and at the same time provide guidance on increasing added value for their agricultural and forest products. In addition, they can assist in the export of products, thus helping to provide foreign exchange for the home country.

Green trade organizations are united in various alliances according to their main focus (ecological, social or economic):

- The International Federation for Alternative Trade (IFAT) brings together producers of handicrafts and food products from developing countries with buyers and managers of "alternative" trading organizations to do business in a way that is beneficial and fair, avoiding traditional intermediaries.

- The Fair Trade Federation (FTF) is an association of wholesalers, retailers and producers committed to providing fair wages and good employment opportunities to economically disadvantaged artisans and farmers worldwide. It links low-income producers with consumer markets and educates consumers about the importance of purchasing fairly traded products which support living wages and safe and healthy conditions for workers in developing countries. FTF also acts as a clearinghouse for information on fair trade and provides resources and networking opportunities for its members.
- The Ethical Trading Initiative (ETI) is an alliance of companies, non-governmental organizations (NGOs) and trade union organizations working to promote correct implementation of codes of labour practice. Its activities include the monitoring and independent verification of the observance of code provisions.

ORGANIZATIONS WITH AN EMPHASIS ON NWFPS

In recent years green trade organizations have been established for NWFP marketing all around the globe. EcoMarket International has been established on Internet as the European platform for green products, services and information. In the Netherlands, ProFound has initiated a NWFP exporters database to provide all potential exporters of NWFPs a means of advertising on the Internet. The Natural Resources Institute in the United Kingdom carries out work comparing ethical and conventional trade in NWFPs and has established the UK Consultative Group on Ethical Trade and Forests.

The following examples of green trade organizations give an idea of some of the different types of organization that place specific emphasis on NWFPs.

Green Trade Net

The Green Trade Net initiative of the German Agency for Technical Cooperation (GTZ) offers assistance for the full range of activities from harvesting of the raw material to processing, research and trade contacts. The Green Trade Net office provides detailed company, product and country information for importers, producers and consultants. The information base currently covers more than 650 organic products (including many edible NWFPs) from 27 developing countries. The following services are available:

- Green Trade Net provides the opportunity for producers to contact importers and to offer their products through the Web site;
- Importers can search for export products via the Green Trade Net office by fax, phone or e-mail;
- Information is available with regard to different product groups such as essential oils, herbs or tea;
- Importers are sent information about specifications, availability, certification status and producers;

- Monthly mailings of current offers in different product groups are sent to importers;
- Green Trade Net distributes product samples to interested importers (of which more than 160 are in contact):
- Direct links are established between registered exporters and importers that have expressed an interest in their product.

The registration of producers and product promotion is done on the basis of written consent. Information given by Green Trade Net is free of charge.

ForesTrade

ForesTrade, a United States-based company specialized in the production and marketing of fine-quality certified organic and sustainably harvested products, acts as a catalyst for the building of mutually supportive alliances of producers, processors and support organizations in the producing countries and as the primary network to the international market. ForesTrade supports local institutional development, training and extension, certification and ongoing monitoring, supervision and quality control activities in partner countries such as Guatemala, for example. Since the United States (unlike other countries such as Germany, the Netherlands and Switzerland) gives virtually no government subsidies for such activities, ForesTrade, a partly privately owned company, must mobilize resources to set up its projects. Therefore, it serves primarily as an international marketing and trading agent for producers in order to recover the costs of the projects.

ForesTrade offers a broad selection and a reliable supply of tropical spices and selected essential oils to natural food processors, distributors, manufacturers and the personal care industry. The products, which include vanilla, cassia, pepper, cloves, nutmeg, mace, ginger, turmeric, allspice, cardamom, coffee and patchouli oil, are primarily produced by a cooperative network of small farmers, indigenous organizations and local businesses in South and Southeast Asia and Central America. ForesTrade provides stimulus and support to environmentally sound agriculture, local economic advancement and rain forest conservation.

The BIOTRADE Initiative

Many developing countries are endowed with rich and highly diverse biological resources which provide a wide range of products and services such as watershed protection, ecotourism and products for, among others, the pharmaceutical, cosmetic, food and agrochemical industries. The BIOTRADE Initiative of the United Nations Conference on Trade and Development (UNCTAD) was launched during the third Conference of the Parties of the Convention on Biological Diversity (CBD), held in Buenos Aires, Argentina in November 1996. Its mission is to stimulate investment and trade in biological resources as a means of furthering the three objectives of CBD:

- Promotion of the conservation of biodiversity;
- The sustainable use of the components of biodiversity;
- Fair and equitable sharing of benefits arising from the utilization of biological resources.

The initiative undertakes to fulfil this mission by developing countries' capacity for sustainable use of biodiversity to produce new value-added products and services for national, regional and international markets. To achieve these goals. the BIOTRADE Initiative works to integrate the private sector, governments. local and indigenous communities and other relevant players in a mutual beneficial framework.

The initiative has three complementary components:

- Market research and policy analysis;
- Web services and communications;
- A country programme.

The Biotrade country programme represents the most comprehensive part of the initiative. It analyses opportunities and constraints for the development of a sustainable bioresource industry. To capture opportunities and solve identified problems, the country programme develops proactive strategies focusing on biobusiness development, partnerships. sustainable use, conservation and benefit-sharing incentives. The other two programme components are designed to compile and analyse market data and policy issues and to disseminate the information thus gathered through a Web site, printed publications and briefings. These components form the basis for a more transparent understanding of market dynamics and trends, market barriers, trade and investment flows, property rights regimes and biobusiness development as well as conservation, sustainable use and benefit sharing.

The final goal for national sustainable biobusiness development will be to launch and implement new partnerships integrating the national stakeholders" interests with the commercial objectives of national, regional and international financing sources, the consulting sector (NGOs and others) and other global players.

THE CHALLENGES OF EXTENSION FOR NON-WOOD FOREST PRODUCTS

Interest in non-wood forest products (NWFPs) has risen in the past 15 years or so as foresters, conservationists and development workers have struggled with the issue of promoting economic returns from forests while simultaneously conserving them. An underlying assumption of the interest in NWFPs is that sustainable economic harvest of "secondary" products will help avoid forest conversion to other economic land uses or extensive logging. However, this approach has recently been recognized as somewhat simplified; accumulated experience, including evidence of unsustainable extraction and negative socio-economic impacts, has led to a more complex view. More consideration is being given to the role of NWFPs in improving rural livelihoods and to the assistance that could or should be provided to help achieve this. As J.E.M. Arnold states in a review of trends in community forestry (manuscript in preparation, FAO Community Forestry Unit): "It is therefore often necessary to be able to distinguish between those forest products activities that feature in the survival strategies of the very poor, and those product activities that can contribute to increasing the incomes of house holds operating in a more dynamic economic environment. This can be very important in determining what support and intervention measures may be appropriate."

THE APPARENT PARADOX

Although forestry extension has been defined in many ways, it may be considered as a systematic process of exchange of ideas, knowledge, techniques and information leading to mutual changes in attitudes, knowledge, values and practices aimed at improved forest and tree management and rural development (Anderson and Farrington, 1996). In organizational and operational terms, many extension systems, public and private, are based on some degree of standardization and economy of scale. Traditionally, extension has often aimed at optimizing the production or yield of a single product or a limited array of goods and services.

For example, because of strong demand, large areas may be managed for fibre production using certain standardized techniques and practices. The green revolution and "training and visit" types of extension tended to view the purpose of extension as the delivery of standard technical packages developed by formal researchers to farmers in order to increase production over large areas of fairly homogenous farmland.

NWFPs, on the other hand, are often low in value, multiple and diverse. They vary greatly over time and space, and they can fluctuate strongly in response to markets. NWFPs can occupy overlapping and competing niches. Fruits, bark and leaves of the same species may provide different products and have different uses. The aim to optimize the sustainable production of an array of products greatly complicates extension approaches. Many NWFPs have not been formally researched to a significant degree and are relatively little known from technical, economic, social and environmental perspectives.

Extension for NWFPs therefore can be seen as something of a paradox: how can systematic and standardized extension approaches deal with goods and services characterized by diversity and multiplicity? NWFPs are growing in importance as income-generating products in many areas. How can the apparent contradiction be addressed so that they can be appropriately promoted?

EMERGING APPROACHES TO EXTENSION FOR NWFPS

Several complementary attempts have recently been made to address the extension challenge. The approaches, which are still being developed and refined, can be characterized under the general headings of socio-economic approaches, integrating local knowledge and networking.

Socio-economic Approaches

In part as a response to the conservation and development emphasis, recent extension efforts have accentuated the need for income generation from NWFPs. This emphasis has led in turn to a focus on capacity building and small enterprise development. Lecup *et al.* (1998) mentioned three approaches: business planning, enterprise and marketing development and market analysis and development. The last is particularly interesting because it integrates social. technical and ecological as well as economic and financial considerations.

Market analysis and development (MA&D) is a methodology for helping entrepreneurs working with trees and forest products to identify products and develop markets that can provide them with income and benefits without degrading the resource base. The MA&D process consists of a series of phases focusing on capacity building and strengthening of institutions at the local level to provide the support that local entrepreneurs need to develop and run small enterprises. One of the goals of MA&D is to enable tree and forest product entrepreneurs to develop and operate their enterprises independently in the long term. The entrepreneur's role is central to the methodology; entrepreneurs are the main actors and decision-makers, even though they may need the initial support of a facilitator, e.g. a forest extensionist.

The methodology enables field staff, facilitators and planners to integrate social and resource management issues in their support to entrepreneurs. MA&D helps the facilitator and the entrepreneur to take a multifaceted approach to investigating the market environment with a view to avoiding potential failure.

Government agencies and programmes and development organizations also benefit from the MA&D approach. The methodology reduces the risk of time and funds being spent on unsuccessful enterprise development. MA&D is a cost-effective process which assists in the development of viable enterprises.

Integrating Local Knowledge

Since local people and groups often know more about specific NWFPs than many formal scientific institutions, integration of local knowledge with institutional research results improves the effectiveness of extension. Participatory research and extension help to ensure that the empirical local knowledge built over many years of living with the forest is captured and used. In addition, scientific methods can be applied to systematize and expand local knowledge. Participatory extension can help build methodological capacity at the local level for critical analysis of trends and action. The combination of the knowledge resources of the various parties facilitates innovation. A participatory study on the role of NWFPs in a conservation and development strategy in Pará, the easternmost state of the Brazilian Amazon, provides an example

(Shanley, 1999). Attempts were made to make the data collected as accessible as possible to local people, for instance through the use of or conversion to local units of measurement, through workshops with local people and through song and theatre, in order to facilitate a dialogue with local forest users. This effort also served to break down the barriers between research and users (including extensionists). The extension team developed illustrated booklets presenting the ecological and market data, posters, songs and lore used in workshops, which could be used to reinforce outreach efforts, could serve as a training tool for extensionists and could reach distant communities beyond the team's travel. Shanley concluded: "We need to question... who the primary beneficiaries of our research really are, as well as the common assumption that our research is complete once the scientific article has been sent to press. Rural extension is an underutilized, cost-effective way to ensure that hard-won field data not only land on the desks of other scientists but are also given back to the forest-based community who need [them] most."

Networking

Another approach has been to emphasize networking, especially between technicians working on NWFPs and potential producers and markets. Since technical knowledge about an individual NWFP may be held by disparate partners, technical networking may be an effective way of combining and systematizing technical knowledge and making it more widely available. An example is the work carried out by FAO's NWFP programme.

Networks are usually viewed as a flexible and often informal means of improving information exchange. Although they may cover wide areas of expertise. NWFP networks often focus on rather specific topics.

With the advent of Internet facilities and e-mail, which allow simultaneous contact with a large number of people at low cost, the capacity and efficiency of some networks have been greatly enhanced. Electronic newsletters (e.g. ntfp-biocultural digest, an e-mail list for announcements and facilitated discussions on topics related to non-timber forest products [NTFPs]; Tropenbos NTFP mailing list) and e-mail conferences are effective and cheap tools for exchange of views and animated discussions on specific topics.

WHAT DOES THE CHALLENGE MEAN FOR EXTENSION?

The problem of developing systematic extension approaches for the numerous and diverse NWFPs strikes at the heart of the question of what extension is and what it does. The development of approaches for cases that initially appear to be difficult, such as NWFPs that are not traded internationally, can inform and enhance extension for more traditional forest products such as timber and fuelwood. Many of these products share characteristics with NWFPs: their value fluctuates over time and space, and local knowledge systems are often essential to understanding issues regarding their sustainable management.

As diversification and differentiation of markets continue, approaches developed for NWFPs may well be important for other products in the future.

STRENGTHENING OF JOINT FOREST MANAGEMENT

As per the provisions of National Forest Policy 1988, the Government of India, vide letter NO.6.21/89-PP dated 1st June, 1990, outlined and conveyed to State Governments a framework for creating massive people's movement through involvement of village committees for the protection, regeneration and development of degraded forest lands. This gave impetus to the participation of stakeholders in the management of degraded forests situated in the vicinity of villages.

The joint forest management programme in the country is structured on the broad framework provided by the guidelines issued by the Ministry. So far, during the last ten years, 27 State Governments have adopted resolutions for implementing the JFM programme in their respective states. As on 15.8.2001, 14254845.95 ha of forests lands are being managed under JFM programme through 62890 committees.

The JFM programme in the country was reviewed by Government of India from time to time in consultation with State Governments, NGO's and other stakeholders in view of several emerging issues. In order to further strengthen the programme, the State Governments may take action on the following suggested lines.

LEGAL BACKUP TO THE JFM COMMITTEES

(i) At present, the JFM committees are being registered under different names in various States as per the provisions contained in the resolutions. Except in a few States where the committees are registered under the relevant acts in most of the states there is no legal back up for these committees. It is therefore, necessary that all the State Governments register the JFM or village committees under the Societies Registration Act, 1860 to provide them with legal back up. This may be completed by 31st March, 2000. Completion of such formation of existing JFM committees may please be reported to this Ministry.

(ii) There are different nomenclatures for the JFM committees in different States. It would be better if these committees are known uniformly as JFM committees (JFMC) in all the states. Memorandum of Understanding, with clearly defined roles and responsibilities for different work or areas should be separately assigned and signed between the State Governments and the committees. All adults of the village should be eligible to become members of the JFM Committees.

PARTICIPATION OF WOMEN IN THE JFM PROGRAMME

Considering the immense potential and genuine need for women's participation in JFM programme, following guidelines are suggested for ensuring meaningful participation of women in JFM.

(i) Atleast 50% members of the JFM general body should be women. For the general body meeting, the presence of atleast 50% women members should be a prerequisit for holding the general body meeting.

(ii) Atleast 33% of the membership in the JFM Executive Committee/ Management Committee should be filled from amongst the women members. The quorum for holding meeting of such Executive/ Management Committee should be one-third of women executive members or a minimum of one whichever is more. One of the posts of office bearer i.e. President/ Vice-President/ Secretary should be filled by a women members of the Committee.

EXTENSION OF JFM IN GOOD FOREST AREAS

For better resource planning and collective management distance from the village and dependency on forests should be the main criteria for allowing JFM programme to operate. Therefore, JFM programme should cover both the degraded as well as good forests (except the protected area network). The microplan or treatment plan and memorandum on understanding should be different for degraded forests and good forests (crown density above 40%). In good forest areas, the JFM activities would concentrate on NTFP management and no alternation should be permitted in the basic silvicultural prescription prescribed in the Working Plan but to promote regeneration, development and sustainable harvesting of NTFP which can be given free or on concessional rates as per existing practice in degraded areas under JFM. The benefit sharing mechanism will also be different for the good forest areas.

The JFM committees will be eligible for benefit sharing for timber, only if they have satisfactorily protected the good forests for a minimum period of at least 10 years and the sharing percentage should be kept limited to a maximum of 20% of the revenue from the final harvest. The felling of trees and harvesting of timber will be as per the provisions of the working plan. A certain percentage of revenue from final harvest should be ploughed back in the silviculture & management of the forests. The extent of good forest areas to be allowed will depend upon the number of village household and should be restricted to a maximum limit of 100 ha and generally limited to 2 km from the village boundary. For degraded forests also as far as possible JFM should be first concentrated on areas upto 5 km from the village boundary. The implementation of JFM in good forest areas shall be done in a phased manner on pilot basis. The pilot areas may be monitored closely for a few years and based on the feedback and success achieved the programme can be extended further in consultation with the Central Government. Before allowing the good forests on pilot basis, all the degraded forests of that locality should be covered simultaneously.

PREPARATION OF MICROPLAN IN JFM AREAS:

(i) In case of new working plans a JFM overlapping working circle should be provided to incorporate broad provisions for micro plans. To achieve this flexible guidelines should be evolved for preparation of local need based micro plans. For this purpose, the working plan officer will work in tandem with the territorial DFO and CF for finalisation of the prescriptions of the JFM overlapping working circle. The micro plans should be prepared by the Forest Officers and Village Forest Protection Committees after detailed PRA exercise and should reflect the consumption and livelihood needs of the local communities as well as provisions for meeting the same sustainably. It should utilize locally available knowledge as well as aim to strengthen the local institutions. It should also take into account marketing linkages for better return of NTFPs to the gatherers and should also reflect the needs of local industries/ markets. This should be done with due regards to the environmental functions and productive potentials of the forests and their carrying capacity as also their conservation and biodiversity values.

(ii) In areas where the existing working plans are in force (till their revision in future), for incorporation of micro plans in the working plans, a special order may be issued by the PCCFs for implementation of the microplan. In these areas, micro plan should aim at ensuring a multi product and more NTFP oriented approach. Without changing the basic principles of silviculture, deviations may be approved in the existing working plans if necessary. To ensure this, the concerned DFO and CF should dovetail the requirements of micro plans with the working plans.

(iii) The micro plan should also take into consideration and provide suitable advice for areas planted/ to be planted on community lands and other Government lands outside the notified forest areas including in the district council areas of North East.

(iv) Infrastructure/ Ecodevelopment under micro plan should form a separate entity for funding it through concerned development agencies.

CONFLICT RESOLUTION

In order to resolve conflicts in the functioning of JFM committees and to maintain harmony among different groups participating in the JFM, State Governments may constitute divisional and state level representative forums or working groups. This forum/ group should include representatives from all the stakeholders including NGOs. The model prescribed by the Andhra Pradesh Government for this purpose is a case in point for consideration.

Recognition of Self-initiated groups

The community groups in many places in Orissa, Bihar, Gujarat, Andhra Pradesh and Karnataka are performing the essential functions of forest protection and regeneration. These groups need to be identified, recognized and registered as JFM Committees after proper verification of records and enquiry. The period of their existence and duties performed for protection and regeneration should be suitable assessed and proper weightage given to them for deriving benefits under the JFM programme.

Contribution for Regeneration of Resoruces:

For long term sustainability of resources, it is essential that a mechanism is created for ploughing back a certain percentage of the revenue earned from final harvest. For this purpose, no less than 25% of the share of village community should be deposited in the village development fund for meeting the conservation and development from its share of such sales. There should be transparent mechanisms for computation of income for sharing the benefits between different stakeholders.

Monitoring and Evaluation

Concurrent monitoring of progress and performance of this programme should be undertaken at Division and State Level. Evaluation of the programme should be planned at an interval of 3 years and 5 years at Division and State level respectively.

6

Green Space Development and Management in Forest

To develop, implement and coordinate the urban forestry programmes for Metro Manila, DENR created the Urban Forestry Division (UFD) in 1988, now Urban "Forestry and Law Enforcement Division (UFLED), under the Forest Management Services (FMS) of DENR – NCR. The Urban Forestry Section under the UFLED has 2 units: the Planting Stock Unit, which is responsible in planting stock production and distribution, and the Cooperative Planting Unit which is tasked to monitor collaborative UF programmes like LKP, CGP and Adopt – A – Street/Park Programmes. The Urban Forestry Section is also currently implementing the OPLAN SAGIP PUNO programme. The urban forestry/greening programmes of DENR- NCR was conceived and adopted to make Metro Manila into a green metropolis. Among its activities include production of planting stocks; establishment and maintenance of mini – forests; greening of main thoroughfares, side streets and islands; establishment of joint nurseries with LGUs, NGOs, POs, schools; providing technical assistance and training on proper site preparation, choice of species and proper planting and maintenance; harnessing the cooperation and involvement of the public via information campaign; periodic assessment of greening activities; processing and issuance of balling permits, and implementation of Oplan Sagip Puno programme.

At the city or municipal government level, specific urban forestry/greening offices were created under the Mayor's office. Some of these offices are ad – hoc in nature while the others are permanently institutionalized in the city government structure. The greening offices of all the 6 cities have a mandate on green space development and management although in 3 cities (Makati, Pasig, Mandaluyong), they also included cleaning, waste management and pollution monitoring and control as their other mandates. In addition to the greening office, each city also has either a committee or Tasks Force created to enhance active participation of other sectors in the greening and cleaning activities.

The capabilities of these greening offices in terms of manpower, available facilities and financial resources were also assessed by Palijon (2000). He found that: a) majority of these offices felt the need for additional technical staff to fully implement their greening programmes such as horticulturist, forester/ arborist, and landscape architect; b) the greening offices have inadequate facilities, equipment and tools needed for their programmes; and c) all cities, except Makati, have insufficient and unsustainable budget for their greening programmes. In the case of the other mega-cities, the DENR regional office does not have specific urban forestry unit unlike in the NCR. Their urban forestry programme activities are being handled by the Reforestation Section under the Forest Resources Development Division of the Forest Management Services, except in Cagayan de Oro where a focal Programme Unit for Urban Forestry was recently created by the RED attaching such unit directly to the RTD-FMS. At the city government level, the City ENRO, Clean & Green Office, Task Force Clean & Green and Cleaning & Beautification Committee are usually the offices or bodies attached or created under the Mayor's Office responsible for management of urban forestry programmes just like those of the NCR.

PHYSICAL ACCOMPLISHMENTS OF URBAN FORESTRY PROGRAMME

The total number of seedlings planted in the urban greening programme by city/municipality in Metro Manila from 1988 to 2002. The 17 cities and municipalities comprising Metro Manila (NCR) have an aggregate total of 2,212,488 seedlings planted. The highest number of seedlings planted was in Quezon City (0.72 million seedlings or 34 per cent of total), followed by Manila City (0.31 million seedlings or 14 per cent of total). On the other hand, the total population of Metro Manila in 2000 was 9.93 M people.

Based on the statistics, the estimated tree to person ratio is 1:6 or 1:9 assuming 80 per cent or 50 per cent survival of the seedlings planted, respectively. These ratios are short of the 1:4 target ratio set by MPFD. However, the estimated number of trees is most likely an underestimation because the trees already existing prior to 1988 were not accounted or tallied. For the other mega-cities, the attainment of 1:4 ratio can not be ascertained due to absence of data on the number of seedlings planted on different years and the absence of inventory data on the number of trees already existing in these areas.

Accomplishment for Mini Forests and Parks

The minimum number of mini-forests or parks targeted by the MPFD for Metro Manila was 60 for the 8.2 million residents and this was based on the assumption that there should be 1 mini-forest or park per 100,000 – 150,000 residents. Since the present population of Metro Manila (NCR) is about 10M, there should be at least 67 mini-forests or parks already established and maintained. As of 1994, there were already 472 parks established in different parts of Metro Manila, which is more than enough (*i.e.* 7 times more) compared to the target ideal number. Almost 50 per cent of the total number of parks in Metro Manila were established in Quezon City. Between 1996 to 2001, eight more mini-forests were established in Metro Manila covering about 18.20 hectares. Hence, for the entire Metro Manila, a total of 480 mini-forests or parks have already been established as of 2001.

Compound Planting

As envisaged in the MPFD, at least 570 compounds or grounds should have been planted or landscaped by year 2000 in all six priority cities. Since there is no breakdown per city, the target maybe equally allocated at 95 compounds or grounds per city. There is no available information on the number of compounds or grounds planted or landscaped in the different mega-cities. Nevertheless, for Metro Manila, data is available on the number of seedlings planted. For the past 14 years, about 1.1 million seedlings have been planted on different types of compounds mostly in the government offices and schools in Metro Manila.

Greenbelt Development

For Metro Manila, the MPFD targeted 100 km of roadsides planted or greenbelt developed from 1990 – 2000. Since there is also no data available on actual length of roads planted or greenbelt developed, we can use the available data on number of seedlings planted along thoroughfares and streets for extrapolation. There were 673,813 seedlings planted along major thoroughfares and barangay roads of Metro Manila for the past 14 years. Due to limited space along road shoulders, it is assumed that there will be a single line planting on each side of the road. At 2m spacing, there will be 1,000 seedlings required per km of road. Assuming these 673,813 seedlings planted included 50 per cent replanting, it is estimated that about 337 km of roads and thoroughfares should have been planted in Metro Manila. This is 3.4 times more than the target of MPFD.

For Iloilo City, the DENR-FMS Region 6 reported 34 adoptors (government agencies, universities, private companies, NGOs, etc.) involved in Adopt-A-Street/Park Programme since 1994. Of the 34 adoptors, 31 adopted streets spanning 17 km while 3 adopted perimeters of parks covering about 1 km. Since the MPFD's roadside planting target from 1996 to 2000 is 35 km, only about 50 per cent of the target was accomplished. In the case of Cagayan de Oro City, there were 10.5 km roadside planting done in 1995 to 1999 in barangays Balubal and Indahag and along Lumbia Airport Road (DENR-FMS Region 10). On the other hand, the City ENRO conducted 8 km highway tree planting/urban greening in barangays Cugma, Tablon, Bugo and other urban barangays (City ENRO – Cagayan de Oro City). Thus, the DENR and City ENRO's roadside planting efforts fall short (*i.e.* only 46 per cent accomplishment) of the MPFD's 1996-2000 target of 40 km. However, for the year 2001-2005, the MPFD's target of 20 km greenbelt development was already accomplished through the tree and bamboo planting project undertaken in 2002 along 20 km riverbanks in barangays Balulang, Lumbia, Bayanga, Mambuaya and Dansolihon through a contract awarded by the ENRO to MARBEMCO, a people's organization (City ENRO - Cagayan de Oro, 2002).

URBAN FORESTRY DEVELOPMENT AND MANAGEMENT

The urban forest is all of the woody vegetation growing in an urban area, including trees, shrubs, and vines found along city streets, public parks and private property. The City is responsible for managing an urban forest that contains over 40,000 street trees and 6,600 park trees (mowed areas).

There are several policies, programmes and projects issued and implemented for the past several decades which is an indication of continuing concern on the deterioration of urban environment. It is noticeable that new policies and programmes evolve whenever there is a change in administration (a common phenomenon in the Philippines) indicative of lack of continuity of previous initiatives. The major policies and programmes related to urban forestry are chronologically listed below:

- PD 1153 of Pres. Marcos dated 1976 (Tree Planting Decree to support PROFEM)
    - Requires all able-bodied Filipinos 10 years old and above to plant a tree per month for 5 consecutive years.
    - Certificates of planting and survival— requirement for graduation from school, renewal of job appointment and business permit and approval of retirement from service.
    - "Halamanan ng Bayan" launched by MHS to support this programme. It required each city or municipality to put up a nursery, garden and park.
    - Repealed by EO 287 dated July 25, 1987 because of dictatorial provisions and harsh penalty.
- PD 953 of Pres. Marcos dated July 6, 1976 (Greening of Private Lands Including Residential Subdivisions)
    - Requires private landowners to plant trees extending at least 5 m on each side of the rivers/ creeks.
    - Developers or owners of residential subdivisions and commercial/ industrial lots to set aside 30 per cent of total area as open spaces for parks and recreational areas.
    - Penalizes unauthorized cutting, destruction or injury inflicted on naturally-growing or planted trees or vegetations in any public places.

- LOI 1312 of Pres. Marcos dated April 23,1983 (Establishment and Development of Local Government Forest or Tree Parks Throughout the Philippines).
  - Requires each barangay, municipality or city to establish and maintain at least one forest or tree park of considerable size.
  - MNR (now DENR) to allocate public lands for this purpose and to provide technical assistance and seedlings needed.
  - MHS to ensure that establishment of forest or tree parks is included in the land use plan of each barangay, municipality or city.
  - MILG (now DILG) to appropriate funds and implement establishment and maintenance activities.
- Memo Order Nos. 198 and 199 of Pres. Aquino dated November 9, 1988 (Luntiang Kamaynilaan Programme (LKP)/ Hardin ng Bayan Programme).
  - Issued to help insure healthy environment in Metro Manila (MM) and to serve as model programme for other cities/municipalities.
  - Anchored on the "Hardin ng Bayan" concept wherein each city or municipality should have gardens or parks of their own, transforming MM into a garden metropolis with lush vegetations, cool and fresh air like the countryside.
  - Objective- to plant 2 million trees in 2-3 years and achieve a desired 1:4 tree-man ratio.
  - For efficient, effective coordinated implementation, an Inter-Agency Committee (IAC) was formed: Co-chair- DENR and MMA (now MMDA); members- DPWH, DOTC, Metro Police Force, DOT, OPS and PMS.
- Memo Cir. No. 5 of Pres. Ramos dated August 27, 1992 (Clean and Green Programme).
  - Similar to LKP (same IAC composition except MMDA as chair/ lead agency) but wider in scope (not only greening but also cleaning activities)
  - Objective- massive planting (0.5 million trees/year or 2.5 million trees in 5 years from 1993-1997) to achieve the ratio of one tree for every 4 persons.
  - Although focused in MM, CGP has nationwide coverage and encouraging cities and municipalities to join nationwide contest for cleanest and greenest city or town.
- EO No. 113 of Pres. Ramos dated July 22, 1993 (Multi-sectoral Tree Planting Activities in Support of ENR Programmes/ECOREV).
  - Scope / Objective – regreening and rehabilitation of all open and denuded lands of public domain, idle lands, private lands and other

suitable areas (both urban and rural) including rehabilitations of coastal and marine areas.
- DENR to identify, assess and designate suitable area for planting and management and to provide technical assistance to participating agencies.
- LGUs implement the programme in their respective level and set up counterpart funds.
- Private sector participation encouraged via MOA or other appropriate arrangements with DENR.

- EO No. 118 of Pres. Ramos dated August 12, 1993 (Mandating the active participation of all government agencies nationwide in urban greening through an Adopt-A-Street/Park Programme)
  - Objective – greening of streets and parks in urban centers.
  - Requires all government offices and government owned/controlled corporations to adopt a street or park in coordination with concerned LGUs, NGOs and private sector by planting appropriate species and maintaining them for at least 5 years using their own funds/resources and other resources.
  - DENR to manage and coordinate the programme through a designated National Coordinator.
  - Project to be turned over to concerned LGU for maintenance and protection.
- DENR-DILG-DPWH-CSC Joint Memorandum Circular No. 1 dated December 17,1993 (Implementing Guidelines for EO 118-Adopt-A-Stree/Park Programme)
  - Described the roles of each participating agency and outlined the schemes in the identification, selection and adoption of a street or park to be developed.
  - DENR to provide assistance to "adopters" in selecting suitable streets or park sites, in providing necessary planting materials and in monitoring performance.
- OPLAN SAGIP PUNO Programme of FMS-NCR/DENR launched on June 5, 2000.
  - Conceived as a component of "Lets Go Green Programme" of former DENR Secretary Antonio Cerilles.
  - Application of appropriate silvicultural treatments (*e.g.* removal of nails, wires/cables, water sprouts; surgical treatment of injured stem or root) to prolong life span and promote good health and vigour of trees planted in parks and along thoroughfares and streets in MM.
  - Supplemented by public awareness campaign.
  - DENR enters into MOA with participating agencies (*e.g.*

subdivision homeowners association, city/ municipal government, NGOs, etc.)

- DENR's role — conduct inventory and assessment of damaged/ injured trees; undertake appropriate silvicultural treatments; conduct information dissemination and training on tree care and maintenance; provide technical assistance and planting materials to sustain the project.
- LGU's role — provide tree care and maintenance crews to sustain the project; assist DENR in information dissemination on maintenance and protection of trees.

• Proclamation No. 396 of Pres. Arroyo dated June 2, 2003 (Enjoining the active participation of all government agencies including government-owned or controlled corporations, private sector, schools, civil society and citizenry in tree planting activity and declaring June 25,2003 as Philippine Arbor Day).
  - Objectives- to promote multi-sectoral participation in tree planting nationwide; to develop greater awareness on the importance of trees in environment, health and human life.
  - Participating agencies, LGUs, schools, etc. to identify areas to be planted in coordination with agencies which have jurisdiction over such areas *e.g.* DENR in case of public lands, LGUs in areas within their jurisdiction, DND for military lands reservation, DOT for ecotourism areas, etc.
  - DENR, LGUs and schools — to establish and maintain nurseries.
  - Respective participating agency/ instrumentality — to maintain and protect the planted seedlings.
  - DENR — to provide technical assistance to all participants.

In general, the following goals and objectives are common to the Urban Forestry (UF) policies and programmes described above:

• to provide /maintain green, clean and beautiful environment;
• to promote public awareness on the importance of trees (promote environmental consciousness);
• enhance people's participation in the programme;
• promote multi-sectoral collaboration, cooperation and support; and
• in the case of LKP and CGP, the specific objective is to attain a 1:4 tree to person ratio to sustain ecological balance.

As strategy to enhance successful implementation of the project, DENR is usually tasked to provide technical assistance in planting, site and species selection, and maintenance operations, including provision of the planting stocks. Understandably, the DENR is also looked up to as the lead agency when inter-agency collaboration is involved in the programme. On the other hand, the city, municipal and barangay governments, which have jurisdiction over the project site, are usually tasked to maintain and protect the tree parks established and streets planted. They are also required to provide counterpart funds and other resources needed for these projects.

At the end of each programme, there seems to be no serious post – project accounting or evaluation of outputs and accomplishments, including evaluation of success and failures. This may be attributed to the fast rate of turn – over of urban forestry/greening programmes being implemented. Another reason maybe lack of manpower and resources to monitor all the projects. For instance, in the case of Metro Manila, the Urban Forestry and Law Enforcement Division Office of FMS – NCR/DENR only has a small unit (Cooperative Planting Unit) under the Urban Forestry Section which is tasked to do the monitoring activities. Needless to say, the synthesis of lessons learned is an important input for planning and formulation of new programmes (*i.e.* we do not have to " reinvent the wheel" so to speak).

DECENTRALIZATION OF FOREST MANAGEMENT

Although a majority of forests continue to be owned formally by government, the effectiveness of forest governance is increasingly independent of formal ownership. Since neo-liberal ideology in the 1980s and the emanation of the climate change challenges, evidence that the state is failing to effectively manage environmental resources has emerged. Under neo-liberal regimes in the developing countries, the role of the state has diminished and the market forces have increasingly taken over the dominant socio-economic role. Though the critiques of neo-liberal policies have maintained that market forces are not only inappropriate for sustaining the environment, but are in fact a major cause of environmental destruction. Hardin's tragedy of the common (1968) has shown that the people cannot be left to do as they wish with land or environmental resources. Thus, decentralization of management offers an alternative solution to forest governance.

The shifting of natural resource management responsibilities from central to state and local governments, where this is occurring, is usually a part of broader decentralization process. According to Rondinelli and cheema (1983), there are four distinct decentralization options: these are:

(i) Privatization – the transfer of authority from the central government to non-governmental sectors otherwise known as market-based service provision,

(ii) Delegation – centrally nominated local authority,

(iii) Devolution – transfer of power to locally acceptable authority and

(iv) Deconcentration – the redistribution of authority from the central government to field delegations of the central government. The major key to effective decentralization is increased broad-based participation in local-public decision making. In 2000, the World Bank report reveals that local government knows the needs and desires of their constituents better than the national government, while at the same time, it is easier to hold local leaders accountable.

From the study of West African tropical forest, it is argued that the downwardly accountable and/or representative authorities with meaningful discretional powers are the basic institutional element of decentralization that should lead to efficiency, development and equity. This collaborates with the World Bank report in 2000 which says that decentralization should improve resource allocation, efficiency, accountability and equity "by linking the cost and benefit of local services more closely".

Many reasons point to the advocacy of decentralization of forest. (i) Integrated rural development projects often fail because they are top-down project that did not take local people's needs and desire into account. (ii) National government sometimes have legal authority over vast forest area that they cannot control, thus, many protected area project result in increased biodiversity loss and greater social conflict. Within the sphere of forest management, as state earlier, the most effective option of decentralization is "devolution"-the transfer of power to locally accountable authority. However, apprehension about local governments is not unfounded.

They are often short of resources, may be staffed by people with low education and are sometimes captured by local elites who promote clientelist relation rather than democratic participation. Enters and Anderson (1999) point that the result of community-based projects intended to reverse the problems of past central approaches to conservation and development have also been discouraging.

Broadly speaking, the goal of forest conservation has historically not been met when, in contrast with land use changes; driven by demand for food, fuel and profit. It is necessary to recognized and advocate for better forest governance more strongly given the importance of forest in meeting basic human needs in the future and maintaining ecosystem and biodiversity as well as addressing climate change mitigation and adaptation goal. Such advocacy must be coupled with financial incentives for government of developing countries and greater governance role for local government, civil society, private sector and NGOs on behalf of the "communities".

THE CLIMATES OF THE PAST AND PRESENT

The planet on which we live is an oblate spheroid, its equatorial region being warmer than the two polar areas. This temperature gradient, in conjunction with the earth's rotation, sets up a reasonably standard pattern of atmospheric circulation. Since the surface of the earth is neither all land nor all water, but divided into land masses surrounded by even larger areas of ocean, the physical properties of water tend to decrease seasonal contrasts near continental margins relative to those in interior regions. Any enlargement of continental areas would tend to increase the seasonal contrasts, induce steeper barometric gradients and thus heighten the general storminess of their interiors.

Conversely, any decrease in the continental areas, as by a lowering of their general masses with an encroachment of oceanic waters, would tend to bring about an amelioration of climate and a general stabilization of seasonal variations. As the geological record clearly indicates, there have been times when the parts of the continental masses exposed above oceanic waters increased for a while and then shrank, with long periods of relative quiescence between, often accompanied by extensive invasions of the sea.

The periods of increase of the exposed portions of the continents were, on the whole, correlated with periods of mountain building and the development of plateaus. The geological record also indicates that these periods of continental uplift were usually accompanied by intermittent glaciation, at least in the polar regions. Volcanic activity of greater or lesser intensity, especially near the continental margins, also was characteristic of such times.

The fact that we happen to be living in one of these cycles of disturbance focuses our attention on the climatic (and consequent vegetational) features of such periods rather than on the more quiescent and far longer periods that have intervened between them. As Russell has pointed out, man during his entire history has not known what it is to live in a "normal" climate. One may also add that, during, his existence, man has not witnessed what might be termed a "normal" phytogeographic pattern.

If the surface of the earth were uniform there would be three parallel zones of higher-than-average precipitation, one in the equatorial region and the other two in the general regions of the 60° parallels of latitude. There also would be four zones of lower-than-average precipitation, two in the polar regions and two centering about 30° N and S latitude. Because of the rotation of the earth and the interaction of factors resulting from the positions of continental masses and oceans — and their unequal distribution — this theoretical pattern is greatly altered. Furthermore, as Rossby has noted, in the northern hemisphere the much greater percentage of land tends to disrupt the expected pattern more than it is disrupted in the southern hemisphere with its much larger areas of ocean.

There is a close correlation between effective precipitation and vegetation. Hence one may draw certain inferences regarding precipitation from a generalized map of the world's vegetation types. In spite of the heat, the precipitation near the equator is such that few desert areas are found there. Apart from polar wastelands, subhumid and essentially treeless regions tend to develop at the western continental margins at about latitudes 20°-30° and to trend inland in a northeastward direction in the northern hemisphere and southeastward in the southern hemisphere. Although local conditions such as mountain ranges modify these relationships, this, on the whole, is the general pattern.

The belts of heavy precipitation to be expected on a uniform-surfaced world at about 60° would also be turned northeastward on the continents of the northern hemisphere and southeastward in the southern hemisphere. In the skewing of these belts they would be depressed about 100 at the western continental margins. Their poleward trends would soon carry them into the polar subhumid areas; thus they would disappear as belts and become more or less regional coastal areas centering at about latitude 50°. Evidences of these are to be found in the northern hemisphere in the areas of heavy precipitation in North America in the Pacific Northwest and also in northwestern Europe. In the southern hemisphere only South America extends into sufficiently southerly latitudes for the southern superhumid area to be fully evident, as in southern Chile; but here no extended area is found, because of the southern Andes and the small total amount of land area available.

Reference to the vegetation map will indicate that the tropics, which we sometimes think of as characterized by heavy precipitation and an extensive development of rain forest, are by no means uniformly so. The less humid areas just north and south of the equatorial regions restrict the band of equatorial rain forest and other factors shift it and break it into segments.

Those areas, loosely referred to as "the tropics," actually are not characterized by a heavy effective precipitation and dense rain forest; the latter is a forest type of relatively limited extent. The montane rain forests sometimes developing on the windward slopes of mountains in frost-free regions are superficially similar to lowland rain forests, but the two would be distinguished in any detailed analysis of forest types. On the basis of theoretical climatology, those longer periods in the history of the world when its surface was relatively quiescent, with the exposed portions of the continents smaller than at present and also with somewhat subdued relief, were times of "normal" climate. Vegetational zonation poleward from the equator was evident, but without the sharp gradients of today. The air may have been slightly more humid than it now is over large areas, but the total rainfall probably was somewhat less.

The latter fact would tend to decrease the extent of the expected "standard" areas of superhumid climate, which were probably little better developed than they now are. Contrarily, with a more uniform air circulation, the subhumid zones entering the western margins of the continents, while in evidence, probably did not give rise to desert conditions except in the interiors of the largest of the land masses. Again on the basis of climatological considerations, it can be postulated that, during periods of world-wide continental depression and quiescence, the frost-free area of the world extended much farther poleward than at present. Theoretically the frost-free area would have reached to the neighborhood of latitudes 60° N and S; actually, judging from plant remains, it may have extended somewhat farther poleward on occasion. However, where one places the frost-free limits in relation to the 60° parallels depends to a certain extent on whether one does or does not believe in the theory of continental displacement.

Lastly, on purely theoretical grounds, the polar regions of such periods would be characterized by a cool but not cold climate. Snow probably fell in winter, but along the margin of any polar sea which existed the climate is not likely to have been much more severe than it now is in southwestern British Columbia, northeastern Nova Scotia, northern Scotland, or southern New Zealand. The abundant plant fossils of both the Arctic and Antarctic regions offer ample testimony that these theoretical speculations are not far wrong; if theory errs, the plants indicate that it may be on the side of conservatism.

TERMS OF SUSTAINABLE LAND MANAGEMENT

SLM can be defined as "the use of land resources, including soils, water, animals and plants, for the production of goods to meet changing human needs, while simultaneously ensuring the long-term productive potential of these resources and the maintenance of their environmental functions" (UN Earth Summit, 1992). In a wider context the term sustainable land management is used in soil and environmental protection, in preservation of ecosystem services and mineral extraction. Sustainable is also used in property and estate management as well as regional planning. Strictly speaking (in a wider /opposite context) the term "sustainable land management" is applied in forestry (this is where it comes from originally), agriculture, land surveying and in combination with land use management.

INCONSISTENT APPLICATION

In the international context the term sustainable is often used in the context of development services. In the course of their work organisations such as the United Nations (incl. sub-organisations and programmes *e.g.* UNEP / UNDP, FAO), World Bank, European Commission and development organisations make use of the term. In the centre of attention is the sustainable use of resources and issues around soil protection. The World Bank defines Sustainable Land Management as a process in a charged environment between environmental protection and the guarantee claim of ecosystem services on the one hand. On the other hand it is about productivity of agriculture and forestry with respect to demographic growth and increasing pressure in land use.

"SLM is defined as a knowledge-based procedure that helps integrate land, water, biodiversity, and environmental management (including input and output externalities) to meet rising food and fibre demands while sustaining ecosystem services and livelihoods. SLM is necessary to meet the requirements of a growing population. Improper land management can lead to land degradation and a significant reduction in the productive and service (biodiversity niches, hydrology, carbon sequestration) functions of watersheds and landscapes."

The United Nations Economic Commission for Europe (UNECE) applies the term in a much wider context. Besides agriculture and forestry they include the mineral extraction sector, property and estate management.

"Land management is the process by which the resources of land are put to good effect. It covers all activities concerned with the management of land as a resource both from an environmental and from an economic perspective. It can include farming, mineral extraction, property and estate management, and the physical planning of towns and the countryside."

In the course of national politics and programmes only few European states use the terminology "sustainable land management". Here Australia and New Zealand are to be mentioned as both countries have agreed on sustainable land management with respect to climate change as part of their government programmes.

In the European context the definition of the European Network for Land Use Management for Sustainable European Cities (LUMASEC) may be used as a reference. It emphasizes the inter- and transdisciplinary cooperation on sustainable land management:

"As management is the human activity meaning the action of people working together in the aim to accomplish desired goals, land use management is a process of managing use and development of land, in which spatial, sector-oriented and temporary aspects of urban policy are coordinated. Resources of land are used for different purposes, which may produce conflicts and competitions, and land use management has to see those purposes in an integrated way. Therefore, land management covers the debate about norms and visions driving the policy-making, sector-based planning both in the strategic and more operative time spans, spatial integration of sectoral issues, decision-making, budgeting, implementation of plans and decisions and the monitoring of results and evaluation of impacts."

Research on Sustainable Land Management

Since 2010 various regional projects from science and praxis work on several topics regarding the sustainable use of land and its management. The research projects are embedded in the international research programme "sustainable land management" which is funded by the German Federal Ministry of Education and Research (BMBF).

As an example, a research on land management is carried out by a research team gathering researchers from Zambia, Lancaster and Birmingham universities. The different researchers are experts in biology, social behaviour, land management,...The aim of the research is to help people in Zambia to have a good use of the Jatropha curcas tree using seeds and its efficiency to be a repellent for animals. It is a sort of sustainable land management. The project is called BKS: Bridging Knowledge system for pro poor management ecosystem services. Land is a very valuable but limited resource, which in the future needs to be used in an intelligent way. Thus on the one hand the aim of the programme is to understand the interaction between land management, climate change and ecosystem services. On the other hand innovative system solutions will be developed to encounter the current challenges of global change and the accompanying land use conflicts.

URBAN PLANNING OF FOREST LAND POLICY

Since urban planning in India is largely concerned with development of land, it would be relevant to briefly consider how perceptions about land and real estate property have evolved. The Indian Constitution initially recognised 'to acquire, hold and dispose of property' as a fundamental right. Consequently when land was to be compulsorily acquired 'compensation' at market price was payable. Subsequently the term compensation was replaced by the term 'amount'. This ideology culminated in the enactment of Urban Land (Ceiling and Regulation) Act 1976 that attempted nationalisation of vacant urban land by paying nominal amount. Finally the fundamental right to property was deleted from the Constitution. The first articulation of the Urban Land Policy was proposed by the Urban Land Policy Committee (Ministry of Health) appointed by the Government of India in 1965. The Committee articulated the following Land Policy Objectives

1. To achieve optimum social use of urban land;
2. To make land available in adequate quantity, at right time and for reasonable prices to both public authorities and individuals;
3. To encourage cooperative community effort and bona fide individual builders in the field of land development, housing and construction;
4. To prevent concentration of land ownership in a few private hands and especially to safeguard the interests of the poor and under-privileged sections of the urban society.

Further the Committee observed that to realise the objectives "there is no escape from large scale public acquisition if the question of guiding urban development or the provision of adequate housing and other facilities is to be tackled effectively and large scale advance acquisition of land would really be in the interests of the society as a whole. It is by far the best and perhaps the only way to put an end to speculation in land and to capture subsequent increases in land values.

These surpluses, where realised by the public authorities, should benefit the community in more ways than one." Not surprising the role models of Indian Town Planners — Delhi Master Plan, Chandigarh, Gandhinagar and Navi Mumbai were all based on public ownership of land. Whether public ownership in fact achieved the land policy objectives in such cases may be a matter of debate. But a verdict on Delhi experience was;

1. It has not been possible for DDA to provide land at affordable prices to low income beneficiaries resulting in large scale jhuggi jhopadi colonies.
2. In the absence of price signals land has been sub optimally used, resulting in over provision to powerful groups, and
3. DDA's policy to auction very few plots at a time and treating the maximum price quoted in such biding as the real market price has in fact meant artificially increasing the land price through deliberate scarcity."

However, securing large-scale public ownership of land implied compulsory acquisition of land. There was considerable discontent amongst the original landowners about the manner in which compensation was determined and paid. The Land Acquisition Act 1894 initially provided the date of declaration of intention to acquire the land as the reference date for determining the market value.

However no time limit was laid down for actual payment of compensation. 1984 amendments introduced the time limit of three years and also provided for payment of interest from the date of award to actual payment or possession of land and solatium of 30 per cent of market value. However the market value is to be reckoned at current use value at the exclusion of expected rise in value on account of future use. The proposed changes in the LA Act and the R and R Policy attempt to remove many of these lacunae. But planned urban development is not being recognised as a public purpose for which powers of eminent domain could be used and in practical terms the proposed method of deciding compensation and rehabilitation package would make recourse to compulsory acquisition of land expensive for lands that also require substantial investment in trunk infrastructure. This would compel search for new paradigm in respect of urban land. The thinkers in the first world too were enamoured by the socialistic notion of community ownership of land.

However on practical considerations they sought solutions short of nationalisation of urban land. The two extreme proposals had one thing in common. Both considered right to own land and right to develop (or build upon) as separate rights. Henry George argued that a private landowner may have right to own both land and development rights. But he has no rights on the rents accruing to land, as they are results of monopoly and not efforts of the owner. He therefore argued that the state has legitimate right to recover 100 per cent of such rents by way of taxes. He allowed the owner to retain the returns on his investments in improvements. He was also prophetic about the ills of public ownership of land.

In 1879 he stated, "I do not propose to purchase or confiscate private property in land. It is not necessary to confiscate land — only to confiscate rent. Taking rent for public use does not require that the state lease land; that would risk favouritism, collusion, and corruption." On the other hand a committee under the chairmanship of Justice Uthwatt in UK suggested that the betterment occurring on account of development of land should be balanced with the compensation to be paid for acquiring land for public purposes. To enable recovery of betterment the Committee proposed nationalisation of development rights (by paying compensation).

Development required planning permission subject to payment of betterment. Despite three attempts to recover betterment since 1947 it has still not succeeded. A cryptic comment on this reads, "The state expropriates property rights, and then charges those from whom it has taken those rights for granting permission to use them on its terms. A betterment levy is wrong in principle, and like most things that are fundamentally wrong, it will always fail in practice" The Indian urban planning thought under the pre 1991 macroeconomic framework was oblivious of property rights and resultant land and real estate market. Hernando de Sotto argued the importance of clarity of property rights and labelled poorly recorded property rights as the 'dead capital' unable to ensure finance for poor. But his arguments have evoked little debate in India. Lack of conceptual clarity about land and property rights have given rise to many expedient policy initiatives.

Instrument of Nationalising Development Rights

FAR essentially a zoning tool in US cities rationalises the intensity of development that could be permitted considering the existing level of development, accessibility and use. The FAR in US cities varies considerable from less than one to 15. However in most Indian cities such considerations are not used in defining FAR. In many states common FSI values are prescribed across all cities as part of state wide building regulations and do not form part of master plan of individual city.

There has been two-fold argument justifying this position. First, varying FSI within a city would be seen as discriminatory between different land owners and second, varying FSI across cities would lead to demand from political quarters to increase rationally defined FSI to an arbitrary level proposed in another city. The general tendency has been to prescribe low uniform FSI (around one). This has meant scarcity of development rights particularly in cities that experience faster economic and population growth and resultant increase in demand for per capita floor space. Instead of adopting measures to reduce scarcity of development rights by rationalising FSI pattern, retaining existing low FSI regime is being implicitly used as an instrument of nationalising development rights beyond prescribed FSI (without paying compensation).

Armed with such nationalised development rights the state administrators could allot these rights on conditions of payment or fulfilling other obligations like providing free houses to slum dwellers. (Hyderabad Master Plan proposed a base FSI and permissible increase subject to payment. Maharashtra Government increased FSI in Mumbai subject to payment that would be equally shared between Municipal Corporation and the state government. Extra FSI is similarly allowed in Chennai. Similarly extra FSI is allowed for rehabilitating existing occupants of rent controlled buildings and slums, free of cost). In this context the objectives of raising revenue or helping a class (not necessarily poor) will succeed when base FSI is low and scarcity of development rights is created. Creating scarcity is not a healthy way of managing any market.

FINDINGS AND SUGGESTIONS OF URBAN PLANNING

Urban planning is basically concerned with the location, intensity and amount of land development required for various space using functions of city life industry, wholesaling, business, housing, recreation, education, religious and cultural activities of the people. Thus urban planning is itself neutral. But the institutionalisation of planning practice within a complex bureaucracy has contributed to the re-politicisation of urban planning. It has become a mode of intervention that is only implemented when it serves the specific interests of the interest group parties.

Thus slum demolition is a process in urban renewal, a process in urban planning, which has become a mode or tool of the ruling classes for fulfilling their own interests. In shaping the city in such a way that it conforms to the upper class notion of the city, it is its interests, which are furthered. In actuality clearance is necessary to make the centrally located areas available to the capitalists for various economic activities.

As far as Indian urban policies has been concerned it has been seen that there has been a lag between the promises made in different plans in paper and actual practical work that has been done. Though the state within its set structure of the society has tried to work for the poor through different urban planning policies and housing policies like the different slum development policies in Mumbai, urban land (ceiling and regulation) act, more peoples participation in development activities through 74th constitutional amendments, different poverty alleviation programmes but in actual reality very less has been done. The politicians, builders, businessmen, slumlords, the elite class has continuously manipulated the different policies according to there need and profit maximisation motivations and left the urban poor in constant misery.

Urban policies in general and urban planning in particular has been become an instrument in the hand of the capitalists to fulfil there needs. Thus there is need to make urban development pro-poor or to evolve an urban development framework to agree to the needs of the most vulnerable sections of the urban population.

They should not be concerned about the needs of the capital alone. This can be done through two things. One is the representation in the planning and policy-making bodies and other is the creation of mechanisms or forums for participation in policy making. Thus decentralization of decisionmaking should be done as has been done through 74th constitutional amendments.

But the decentralization should mean change in the structure or power sharing in the society and not decentralization of some convenient functions or responsibilities in a centralized society as of now where economic processes and decision making paradigms are centralized. So decentralization is not only about taking it down to the community but looking at how communities themselves can determine as stakeholders what should happen around them and whatever happens around them should be for them.

People's participation, decentralization, and Privatisation should not be taken as synonyms of each other. Role of the state and role of private sector and people's needs should be appropriately and adequately discussed so that some general understanding is arrived at, as there cannot be a single model applicable to all the urban housing and basic services and utilities. All dimensions, political as well as economic, of the 74th constitutional amendment be analysed and understood realistically and not on ideological or emotional basis. A proper understanding of urban institutions and their functioning has to be developed. The outcomes of NGO's acting as catalysts in the community actions towards development are very encouraging. The experiences and experiments have been regarding alternative institution building, where the stakeholders themselves are directly brought as actors in the development process.

But NGO's as the solution for housing struggle of the urban poor has also been questioned upon. This has been mainly because their internal limitations set to them like the financial constraints and the external limitations like the power politics and administrative constraints. Housing today is looked upon merely in real estate terms. This is what the real –estate agenda has encouraged today due to the Privatisation thrust in housing and corporatisation of the various development and construction activities. Housing projects are evaluated in terms of size, the built up area, the FSI consumed, the financial turnover, and various other business and marketing merits. The bigger the project, the better it is and the greater the attraction for developers in undertaking the scheme. A huge network is thus established between the developers, the landowners, and the financial institutions wherein the slum dwellers find no place.

Thus there is need of changing in the attitude of the government and the elite towards the slum dwellers. Programme for slum development must primarily be seen as an environmental scheme and not merely as an agenda for real estate development and construction turnover. It is the slum-like conditions *i.e.* lack of drinking water, inadequate toilet facilities, garbage, heaps, lack of sewage disposal, absence of open-spaces, inadequate and unsafe access that are of primary concern. Even with the rehabilitation programmes by the

Government and the NGO's some basic element is lacking in these projects that is the people. This has been particularly the case of the rehabilitation of the evicted SGNP slum dwellers.

It is obvious from the above argument that neither the state nor the private sector alone can handle the problem of housing for the poor. So there is need for the NGO intervention. But NGO's have there own drawbacks in terms of financial and administrative constraints. So the solution to overcrowding and housing is to ensure that the socioeconomic compulsions that force mass migration towards the cities are ended. However another more rational way out will be the integrated efforts of the private sector, central government, civil society and people. However, this requires simultaneous efforts by national, local governments, civil society, and private sector and people themselves to eliminate impediments at all the levels. While central governments address policy matters and regulatory impediments nationally, local authorities and civil societies should design strategies to make appropriate interventions and regulatory changes in the city. Local experiences should be fed back to national governments to influence their support to cities, as well as for redesigning national programme.

Another thing that can be done is the site and service scheme that has been started by the government in Fourth plan and continued in seventh plan. This is a solution by which the 5 million homeless workers of Mumbai can be housed. In this scheme the government is totally responsible for providing all the services such as water, electricity, sewage, drainage, and toilets, for providing technical know how and skills, interest-free loans and for providing building materials at highly subsidized rates. First of all land is taken over under the Urban Land (Ceiling and Regulation) Act at the rate of Re1/-a square foot which is virtually free or at adequate compensation or if land is already in possession of the government it can be used directly.

Instead of This land is divided into thousands of small plots. Then no construction is actually done in the beginning, as is the case of conventional schemes. But water supply, sewage drainage and electricity are provided to each plot. Then, depending on the financial capacity of each person the worker can build his house. The advantage is even a person with a budget of 2000/-can go ahead. Secondly the house can be developed over time. Thirdly and most important, the worker has control over his house. Housing must be made fundamental right enshrined in the constitution.

Both the availability of direct finance and subsidized building materials will tremendously excite the slum dwellers in under taking the renovation and reconstruction of their houses on their own. From the table given we can see that people with different income group will be able to afford different types of shelter within there limited finances. Like People with income of as low as ₹.200/-can afford a house of cost 3,200 where the site and services has been provided by the government and the person can make a house within his or her budget. Whereas, a person with a monthly income of more than 1500, can afford a house of ₹.56, 000 made in a conventional way.

Also under the development proposal, the house will be ground and ground + one or two stories high, enabling easy repairs and maintenance directly under the control of the users. There would, therefore, be less dependency on hired skills and services.

This will encourage people's participation in decisionmaking and will inculcate a greater sense of belonging resulting in personalization of spaces and structures. While the construction of houses will be the individual's prerogative, the restructuring of the slum-layout, road, services, open spaces etc., could be a collective effort with governmental support

Development programme for each slum will have to be evolved independently and relevant guidelines fixed. Even F.S.I for each slum may vary to enable housing for all the slum-dwellers on, as–is-where–is basis. Each slum has its own peculiar situation and needs. For example what will apply for Dharavi may not be relevant for another slum in Jogeshwari and vice versa. Therefore, within the main policy framework, individual development strategies will have to be evolved. This will encourage people's participation in decisionmaking.

PROMOTE SUSTAINABLE DEVELOPMENT IN FRINGE AREAS

Control of Development on Fringe Areas: In metropolitan cities and mega cities, urban development is mostly in new settlement areas and new activity centres with planned infrastructures and facilities in the fringe areas to accommodate the increasing population and activities.

Unplanned urban sprawl grows around such centres on the agricultural lands taking advantage of the nearby facilities and infrastructure. In such cases, from an environmental perspective, regulations for protecting the agricultural and vacant lands by restricting developments and the stipulations of the regulations may be as below.

- No use, other than agriculture or irrigation facilities, is permitted.
- Existing water bodies to be preserved.
- No new building or extension of any existing building exceeding the height of 3.75 metres shall be allowed subject to the total covered area of 50 sq. m.
- The minimum front and side open spaces shall be 2 metres and the minimum rear open space shall be 5.00 metres.

Redevelopment of Blighted Pockets in the Central Area: Blighted pockets of low key commercial areas are observed within the core areas which, due to the small sizes and multiple land ownerships, do not get redeveloped as per the existing regulations. If these areas can be redeveloped in a planned manner much of the demand for commercial floor spaces can be served. For the redevelopment of such areas incentives towards land assembly by amalgamation of plots and additional floorspace ratios need to be given. The regulations in this respect may be:-

1. For land assembly exceeding one acre in size, the additional floorspace ratio would be 20 per cent above the permissible limit.
2. In case of land assembly of less than one acre the additional floorspace would be 10 per cent above the permissible limit.
3. In such cases the developments should confirm the zoning regulations.

Compulsory Rainwater Harvesting in New Area Development: Private developments in the form of sub-division of mother plots (for plotted developments and for apartments) is taking place in the fringe areas and adjoining municipal and non-municipal areas. There are regulations in most of the plans which vary according to the size of the mother plots and the regulations specify the minimum width of roads, the percentage of open spaces, the land for physical and social infrastructures *viz.* drainage, water supply (pump house and water treatment plants), sewerage (sewage treatment plants or oxidation ponds), school, health centre, market, milk booth, post office, power substation etc.

In order to comply with the present efforts towards utilization of natural resources by rainwater harvesting and ground water recharging, the regulations for subdivision should include the mandatory provision of community pools of sufficient size so that the rain water from the area can be stored in such pools. The water may be supplied to the community for uses such as gardening, car washing etc. Adapting physical planning to promote sustainable development: efficient infrastructure planning.

Prescription of street alignments for regional roads in fringe areas: The regional roads (National Highways, State Highways or District Roads), connecting a city with the hinterland are often constricted in the fringe areas due to lack of scope for widening owing to dense developments and abutting built up areas. To avoid such situations where the regional roads pass through vacant areas in the fringe of the city, advance actions for prescription of street alignments may be made without acquiring land. The proposed right-ofway as per future requirements may be prescribed and the regulations may be as below.

- For any development on the adjoining plots on the regional roads, the owner/developer would have to make a setback following the proposed right-of-way line and the owners/developers would be allowed the same floorspace as they were eligible for the original plot.
- In such cases after the alignment is notified no subdivision of the adjoining plots would be allowed.
- In cases where the plot size is such that no development is possible by allowing the set back, the local body would have to acquire it.

Dispersal facilities around transport nodes: The railway stations, the regional bus terminals/stops are the important transport nodal points of the urban areas. These areas get congested with dense unplanned commercial and residential developments and in course of time become too congested for easy dispersal of passenger and vehicular traffic. Given the urban growth rates future public transport will carry larger volumes of passengers. Therefore, the areas required for dispersal facilities will need to be increased.

Where such areas are already congested, redevelopment plans are to be prepared by the local body and the redevelopment actions may be initiated, including through private-public partnership projects. In less pressured situations regulations for control of development should be made at least for the area within 200 metres on all sides. In such areas provision should be made for adequate parking facilities for different categories of vehicles including the parking and loading/unloading facilities for transit and paratransit vehicles.

Adequate width of the connecting roads, exclusive roads for pedestrians or grade separated pedestrian facilities should be made. In order to achieve these, specific plans should be prepared indicating the future right-of way of the roads, the parking areas for different categories of vehicles, the pedestrian-only roads and the integration with the railway station or the bus terminal. The regulations in such areas may include the following:

- The Floor Area Ratio (FAR) for buildings of different use categories and means of access would be half of the permissible FAR in other areas.
- The compulsory provision of parking spaces for buildings of different use categories would be double of that required in other zones.
- No new cinema halls, theatres and entertainment centres would be allowed within the area.
- For buildings with retail commercial in the ground floor, the minimum front open space would be 5 metres.

Environment protection around relocated hazardous uses: For implementing the development plans, the non-conforming uses located in a scattered way are required to be relocated in the specified areas in the fringe within a stipulated period. Such uses may be the obnoxious and hazardous industries, tanneries etc. Normally, such areas for relocation are selected beyond the city limits and in the vacant and agricultural lands.

When such relocation of activities takes place, unplanned developments adjoining such centres occur and gradually these areas grow.

To avoid further environmental hazards, the relocation areas should be provided with a buffer zone where no developments other than agriculture and pisciculture are permitted. In this regard, regulations of the Pollution Control Board are to be followed. But in the development plans there should be specific regulations in respect of the buffer zones. The regulations in this respect may be:-

- The area covered by 200 meters on all sides from the boundary of such areas would be designated as Buffer Zone.
- No developments other than agriculture, pisciculture and plantation of trees would be allowed within this zone.
- The existing residential uses within the Buffer Zone would have to be relocated within a stipulated period.

Adapting physical planning to promote sustainable development: conservation of water resources and waste management

Protection of Water Fronts: The water fronts (sides of rivers, canals, lakes and big ponds) in many cities are encroached by unauthorized users and developed in an unplanned manner.

These water fronts need to be protected to ensure proper drainage, and access for open-air recreation, water transportation and protection against soil erosion. Area within 100 metres from the banks should be designated as Water Front zone and specific regulations should be prescribed. The regulations for such zone may include:-

- No new building within 30 meters from the edge of the banks would be allowed.
- In the area lying between 30 metres and 100 metres from the edge of the banks no building more than 5.00 metres in height and 30 metres along the waterfront would be allowed. In case of buildings on stilts the maximum height of the buildings shall be 6.50 metres.
- There shall be a linear gap of 50 metres between two buildings alongside the water front.

Environment protection around solid waste disposal sites: Solid waste management is one of the most critical problems of cities.

The locations of the intermediate collection sites and the final disposal grounds need special attention in consideration of the environment hazards of the nearby localities. The intermediate collection sites are generally located within or near the settlements and therefore need to have a buffer zone. This buffer zone should cover at least 30 metres on all sides.

The regulations in this respect may be:-

1. The intermediate collection site may be designated as Inner Disposal Zone.
2. The area should be provided with boundary walls of at least 3 metres high on three sides.
3. The actual dumping area should be circumscribed by two to three rows of trees in the buffer zone.

GREEN INFRASTRUCTURE PLANNING

The EU Green Book ascertains that there are many threats to green areas, for example from vehicles and advertising (Commission of the European

Community 1990). It also stresses the need for an element of nature in the urban environment. Even if urban nature perhaps cannot compare to 'natural' nature, it is still of great importance. The nature that exists in the landscape is the source we should use for disseminating wild plants and animals into urban areas. This is why green corridors leading from the countryside into urban areas are vital, as is the transitional zone between urban and rural areas. In many places, the green structure is not cohesive and thus cannot provide the necessary transport routes. Just a minor interruption of a green corridor can prevent it from functioning. Although binding the green network together in urban areas demands a great effort, it is absolutely essential if we are to succeed in developing urban biotopes and creating the conditions needed for a richer flora and fauna. The EU Green Book recommends that public green space should be increased as much as possible. Hence, green infrastructure planning includes a holistic assessment of the green infrastructure, current conditions and plans for development that could be achieved, for instance, by making it possible for local councils to give a green plan formal status as a part of council planning.

Examples

Germany is one of the countries at the forefront of green infrastructure planning. The German nature and environmental protection legislation also regulates parks and urban green areas. Their green plans are included on several different levels in landscape planning (which is linked to general plans) and in green structure planning (which is linked to local plans). Although these plans are not legally binding they still play a significant role. Examples from three German cities illustrate the divergent ways in which these plans originate and how they are used.

In Hannover, the overall green structure is included in two planning phases. The park administration has the option of conducting a landscape analysis before work is started, in all forms of area utilization. In this way, planning takes landscape potential into account from the very beginning. Plans for the green structure are combined with the plans for building development in the next phase. Both phases are laid before the politicians, who thus gain the opportunity of seeing how consideration is given in the final proposal to the original, overall intentions. In cases where it is impossible to avoid damaging nature, the authorities can demand compensation. Advantage of this is often taken by the city head gardener to demand improvements elsewhere, for instance a builder can be obliged to undertake the improvement of a neighbouring park. In this way, nature - the green area - becomes a commodity that can be traded, for better or worse.

Stuttgart approaches superior green planning another way. This city has a long tradition of securing the expansion and improvement of its green structure through exhibitions held every 10 years, most recently in 1993. Stuttgart also offers a classical example of the fact that it is not only the lack of green areas that can present a problem; the problem can also be related to access to these areas and their lack of mutual cohesion.

The major problem in Stuttgart was the cohesion of the old royal gardens in the centre of the city. For this reason 10 bridges, which linked the parks together across various arterial roads, were built in conjunction with the Bundesgartenschau of 1977. This was a phase in the long-term 'Green U' plan, which was completed in connection with the 1993 exhibition. The Green U now makes it possible to pass through green areas, from the central royal gardens to the forests at the edge of the city.

In Munich, a plan was adopted in 1992 for 14 green strips, totalling 584 ha, which are to link the city's green areas to the surrounding landscape. The green strips are to be established over a 25-year period and will cost a total of $US430 million. The reason for this major effort is the fact that the development of the city's green areas has not increased in proportion to building development. At the same time, increased awareness of the significance of green areas to ecology and recreational activities has meant that greater attention is being paid to the importance of the cohesion of green areas. A large part of the green areas are on the edge of the city and are owned by it. Only the innermost 10-m belt of this resembles a park and consists of small tree plantations, a green and perhaps a stream. It also contains a network of footpaths and cycle paths. The remainder of these areas is rented to farmers, who operate them with only limited use of pesticides.

In Denmark, a successful example of green infrastructure planning is the 'Finger Plan' for the area outside Copenhagen. The name relates to the fact that the overall structure is shaped like a hand, with the five 'fingers' being planned for urban development. The Finger Plan, which was born as long ago as 1947, had the goal of stopping the layered growth of the city and ensuring that urban growth would thereafter be concentrated in narrow urbanized areas ('fingers') along traffic corridors. The wedges between the fingers were to be used for agricultural and recreational purposes. This idea was very clear and was beneficial to the green areas. Subsequent developments have largely adhered to the guidelines of the plan.

In England the planning of the new towns was founded, even as early as 1900, on the concept of using trees and woodland as a setting for urban development. The aims of the new towns were to relieve the urbanization pressure on large cities such as London, Birmingham and Glasgow, and to create self-contained and balanced communities for living in, for employment and for recreation. The new towns were designed and developed by development corporations. The success of the new towns is to be found in the fact that they were produced as integrated structures with good communications, high levels of amenities and higher than average physical environment, based to a large degree on social commitment and the belief that good physical environment was good for people as well as for business.

Conflicts in Management

In the policy-making, planning and development of urban forestry, some general trends and challenges can be recognized. These comprise a growing complexity, an increasing role of public participation and conflict management, and the transition towards closer-to-nature forests. In a study of urban forest planning and policy-making in 16 European cities, the main conflicts were discovered to be conflicts in relation to nature conservation, forestry vs. urban development, conflicts between different types of recreation and recreation vs. conservation management.

THREATS TO GREEN AREAS

Besides the harsh growing conditions that plants experience in the urban environment, there are other, more immediate threats to urban trees and woodlands. These are the pressures of urbanization, which compete with the green areas in demand for land; hence the high recreational pressure on the areas left for urban green spaces.

URBANIZATION PRESSURE

There is great pressure on urban space resources. Even where this is a question of public buildings, such as museums, where green area remains accessible to the public, it would in many cases detract from the overall recreational quality of the area. In Mexico City, the proportion of green areas within the city is falling by about 3.7% annually.

These are often replaced with buildings, especially in the poorest quarters of the city. Traffic installations and noise are other threats to green areas. Roads can isolate green areas from each other, which reduces their recreational value and their value as corridors for the propagation of flora and fauna. The annoyance caused by noise is more indirect. Dutch studies indicate that road noise annoys about 20% of the population, whereas about 11% is annoyed by air traffic. These figures apply to indoor annoyance. Outdoors, people are exposed to even more noise, especially since part of the urban green areas consist of 'residual areas' along traffic constructions.

Social Factors

Another important urban stress factor is vandalism. This is usually perceived to be a serious factor, although surveys shows that it is not: it is usually the result of careless play and affects only the most exposed trees. Vandalism is predominantly a social problem. Successful community landscaping and gardening in densely populated inner-city neighbourhoods have shown that one deterrent to vandalism is the development of a spirit of proprietorship in residents. Nowak *et al.* (1990) noted the highest tree mortality in areas of lower socioeconomic status. Percentage tree mortality was most strongly correlated with percentage unemployment.

Economic Cuts

In addition, economic cuts are an obtrusive threat to green areas. Park administrations responsible for a large part of city green areas have been hit hard in recent years by cuts in appropriations and personnel. As an example it is typical of the trend in Denmark over the past 20 years that economic resources have dropped by 10-20% while green areas have increased by 20-40%. Similar trends have been observed in other countries, for instance Sweden, Germany and the UK. The sector responsible for the planning, establishment and operation of green areas in Denmark (including the administration of natural and recreational areas) has an annual turnover, public and private, of $US1.35 billion and employs the equivalent of 30000 full-time employees. The maintenance costs of urban green areas amount to approximately $US16 per inhabitant. In this respect, if Denmark can be regarded as an average country in European terms, this gives a total European urban greening maintenance cost of approximately $US3.84 billion. In 1991, it was estimated that at least $US300-350 million was spent annually to enhance amenity tree resources in the UK.

ENVIRONMENTAL ASPECTS

Trees influence the urban environment in various ways, whether individual trees or the entire urban forest. Urban trees can mitigate the environmental impacts of urban development by moderating climate, improving air quality, providing habitat for wildlife and reducing noise levels. The environmental benefits associated with urban trees and woodlands are widely recognized. In large cities such as Chicago and Mexico City, the environmental aspects are integrated in the city greening programmes. At the UN Conference on the Environment and Development in Rio de Janeiro in 1992, all participating countries adopted Agenda 21, an action plan that obliges them to work towards sustainable development, an obligation that in turn devolves to the administrators of urban green areas.

LOCAL-SCALE CLIMATE

The replacement of natural surfaces with buildings and roads, as typified by cities, has altered the thermal and moisture properties of the area, which modifies the local atmosphere and generates the 'urban climate', with poorer air quality and increased air temperatures. By transpiring water and shading surfaces, trees lower local air temperatures.

In areas where the summer is hot, trees are important because of the shade they provide. Trees can even reduce energy use in buildings by lowering summertime temperatures, shading buildings during the summer and blocking winter winds and consequently reduce the emission of pollutants from power-generating facilities.

Trees and shrubs can be used for manipulating air movement in order to control the impact of winds, by obstruction, guidance, deflection and filtration. The resulting effects depend on plant size, shape, foliage density and retention, as well as the placement of the plants. Combinations of trees and shrubs provide the most effective barriers, which may even be created by making use of already existing landforms.

Windbreaks provide protection for considerable distances, up to 20 times the average windbreak height. A medium-porous windbreak (40-60% reduction) reduces wind speed nearly as much as an impenetrable barrier. Meanwhile the shelter from this type of windbreak extends over a longer distance downwind, with much less turbulence present. However, the type of optimum windbreak depends on the situation. Hence, in some settings an impenetrable windbreak will provide the most appropriate protection, whereas in others the highly porous windbreak will be the optimal solution. Thus, the structure of windbreaks in terms of suitable provenances and clones of species, number of rows and density can be designed to provide the type of shelter needed.

Air Quality

Trees intercept particulate matter and absorb such gaseous pollutants as O_3, SO_2 and NO_2, thus removing them from the atmosphere. Trees also emit various volatile organic compounds, such as isoprene and monoterpenes, that can contribute to O_3 formation in cities.

Studies conducted in Frankfurt am Main have shown that streets along which trees are planted contained 3000 polluting particles per litre of air, whereas streets that lacked trees had as many as 10 000-12 000 particles per litre of air. Protective plantations along heavily trafficked roads and around industrial areas are therefore an effective means of reducing air pollution. However, this should obviously not be taken as an excuse for neglecting to combat pollution at its source. In The Netherlands, some cities operate a tree-planting plan, the objective of which is to plant as many trees as needed in order to absorb the quantity of CO_2 emitted in providing the city with heating. To attain this goal, support is also given to reforestation programmes in eastern Europe and Africa, as well as to the extensive tree-planting carried out in the cities themselves.

Mexico City's Urban Forest Programme is another example of increasing awareness of the needs of urban greening as a means of altering the overall quality of life in the city. The fundamental goals of the programme are to mitigate the effects of the severe air pollution in the metropolitan area, to establish more green areas within the city and to give the city a more welcoming appearance.

Because trees can store carbon for long periods, they are an important carbon sink. Even though plants absorb CO_2 and produce O_2, it is important not to assign excessive significance to their role in the urban environment. Harris (1992) reminds us that plants really have only a minor effect on the CO_2 and O_2 content of urban air. Photosynthesis in the oceans accounts for 70-90% of the world's total O_2 production, for which reason it is absolutely vital that they be protected against pollution. However, even a minor reduction in the O_2 content of the air will cause a large percentage increase in its CO_2 content, which would reinforce the greenhouse effect, thus leading to a rise in the global temperature. Harris (1992) also stresses that urban vegetation has an especially beneficial effect on air pollution through its ability to reduce the quantity of airborne particles. Studies of 9000 trees in Chicago have shown that the trees reduce air pollution by 12 t of CO_2 and 10.8 t of O_3 per day.

Biodiversity

Green areas play a vital role in urban biodiversity because these are the main habitats of urban plants and animals. The size and location of the urban green areas are of importance for their qualities as biotopes. Vacant urban wasteland in inner cities often possess a high diversity of species adapted to urban environmental conditions and pressures. For instance, older well-established installations attract birds and mammals whose natural habitat is the forest. The urban forest structures are also of great importance, providing wildlife corridors and stepping stones that connect the habitats in the urban green areas. Swedish examples emphasize that only if the urban infrastructure is integrated with the green infrastructure is it possible to achieve biotope protection within cities.

Since a major part of Europe's population lives in urban areas and receives its daily perception of nature therein, nature in urban areas is important for environmental awareness and an understanding of nature.

Urban areas contain more of nature than is immediately apparent. Older gardens and parks, not to mention churchyards, often have noticeably rich biodiversity. Nature created by humans is often considered to be inferior to nature that evolved without human intervention.

In support of this, it is often asserted that, for example, the number of species is often greater in untouched nature. This is mainly related to the fact that many urban green areas do not boast particularly rich biodiversity. Most of them were established with large paved areas, gravelled areas, well-mown greens and isolated individual trees. However, Owen (1992) showed that in an urban private garden with variegated flower beds and a well-proportioned mixture of cultivated and uncultivated plants it was possible to attract a large proportion of the indigenous species of butterflies (34%), moth (30%) and hover flies (36%).

The presence of birds in the immediate environment is a vital element in the context of recreation. Their presence in urban areas depends on the character of the local vegetation. Variation in the structure of plantations also forms a basis for a rich assemblage of birds. As it takes many years for new plantations to mature, it is vital to preserve as much of the existing vegetation as possible when establishing new areas.

Sustainable Urban Forests

Sustainable urban forests are defined by Clark *et al.* (1997) as 'the naturally occurring and planted trees in cities which are managed to provide the inhabitants with a continuing level of economic, social, environmental and ecological benefits today and into the future'. They will provide long-term net environmental, ecological, social and economic benefits (Clark *et al.* 1997; Miller 1997). To the components of a sustainable forest, McPherson (1998) includes features of adequate species and age diversity, a large percentage of healthy trees that are well adapted to local growing conditions, and a climate-appropriate treecover with native forests stands as one component of overall canopy cover.

7

Trees, Plantations and Forest

Trees outside the forest are defined by default, as all trees excluded from the definition of forest and other wooded lands. Trees outside the forest are located on "other lands", mostly on farmlands and built-up areas, both in rural and urban areas. A large number of TOF consist of planted or domesticated trees. TOF include trees in agroforestry systems, orchards and small woodlots. They may grow in meadows, pastoral areas and on farms, or along rivers, canals and roadsides, or in towns, gardens and parks. Some of the land use systems include alley cropping and shifting cultivation, permanent tree cover crops (e.g. coffee, cocoa), windbreaks, hedgerows, home gardens and fruit-tree plantations.

Classification of trees outside the forest presents certain difficulties. There are existing classifications for agroforestry, but none applicable to all trees outside the forest. For practical reasons, the FRA 2000 definition of "forest" combines aspects of both land cover and land use. This approach creates difficulties not only for classification of forest, but also for classification of TOF.

In a study on data-gathering on TOF in Latin America, where classification was primarily based on land use criteria, separating land use and land cover aspects was found to be a main source of misinterpretation. There was a possibility of confounding coffee plantations and trees in pasture with forest, given their high density.

This clearly shows some of the problems involved in establishing a simple and reliable *a posteriori* classification.In France, the National Forest Inventory (IFN) and Teruti Land Use Study-have begun to attempt coordinating classifications of trees outside the forest. The objective is eventually to use the annual Teruti data to update the IFN ten-year data, with a single national nomenclature as a possible end result.

FUNCTIONS AND CHALLENGES

In industrialized countries, farmers list shade and shelter, soil protection and improvement of the landscape and rural environment as their main reasons for growing trees. In the tropics, farmers grow woody species for food security and subsistence. Trees outside the forest are a major source of food. Livestock fodder produced by TOF can be a matter of life and death in semi-arid or mountainous areas.

Fuelwood remains the prime source of energy in developing countries, representing up to 81 per cent of the wood harvest (FAO 1999). In contrast, in the industrialized countries, fuelwood accounts for less than 10 per cent of total fuel consumption (FAO 1998). Very few studies have reported on overall fuelwood output from stands and single trees outside the forest, but agroforestry systems and orchards are known to provide a large part of the resource.

Trees outside the forest have an important ecological role. Planted trees and shrubs in fields help to check runoff and erosion and control flooding, as well as helping to purify water and protect against wind. Trees lining rivers and streams help to maintain biodiversity, providing spawning beds for fish and shellfish and shade which reduces eutrophication.

The unique role of trees in soil protection and conservation, checking wind and water erosion and maintaining soil fertility is universally acknowledged. Also important are the cumulative benefits of trees on smallholdings to soil and water conservation, in particular in the larger context of mountain watershed management; their positive impact on climate; and their role in buffering the effects of desertification and drought.

RESULTS OF SELECTED STUDIES

In spite of the limits of data at the regional or global level, a number of local initiatives have been carried out. The approaches of the various studies differ in accordance with the purpose and scale of the analysis. Few studies use methods resembling the conventional forest inventory. Many studies rely on existing literature or estimates drawn from surveys and interviews. The quantification of products is often based on different parameters, such as estimates of global output, marketed output, observed or potential productivity or economic value. Thus the reliability of the results is uncertain.

The following are the results of some national initiatives that have assessed trees outside the forest. In Kerala, the most densely inhabited state of India, a study estimated that of the total annual production of 14.6 million cubic metres of wood in the state, about 83 per cent was from homesteads (house compounds and farmlands), 10 per cent from estates (plantations of rubber, cardamom, coffee and tea) and only about 7 per cent from forest areas 26.6 per cent of the state area is under forest cover. Trees outside the forest met about 90 per cent of the fuelwood requirements of the state. Fuel from coconut trees alone, including both wood and non-wood materials (pruned and fallen), constituted about 70 per cent of the total fuelwood supply.

A study in Haryana State in India, an intensively cultivated state with about 3.8 per cent of its area classified as forest land but only about 2 per cent under actual forest cover showed that farm forestry (trees along farm bonds and in small patches up to 0.1 ha) accounted for 41.2 per cent of the total growing stock of wood. Multiple tree rows along roads and canals accounted for 13 per cent and 9.6 per cent, respectively; village woodlots for 24 per cent; and block plantations of less than 0.1 ha for 10.6 per cent.

In Morocco, where forest cover is less than 5 per cent of the land cover and other wooded lands only 7 per cent, nearly 20 per cent of the land may be occupied by trees outside the forest, namely as wooded pasture (84 per cent) and fruit-tree plantations (12 per cent) (Rosaceae, citrus, olives trees, palm trees, walnut trees, fig trees, almond trees).

Fruit production has an important place in the national economy. It is noteworthy that even when a forest is largely destroyed, the carob is one of the few species traditionally conserved, as it is highly appreciated by farmers for multiple purposes, providing both fodder and income from the sale of its fruit for export. However, there are no reliable data on the distribution and potential of this "forest" resource, which is of interest for farmers, herders, concessionaires and the government and distributed on agricultural and forest lands.

In the Sudan, the National Forest Inventory has undertaken a national land use inventory to provide area and volume statistics for planning at the subnational and national levels. The inventory was designed to provide preliminary estimates regarding products other than the traditional fuelwood and timber, such as the amount of gum, fruit or nuts that can be collected and the distribution of non-wood species of interest.

In Costa Rica, the Tropical Agricultural Research and Higher Education Centre (CATIE), in collaboration with Freiburg University, Germany, is developing a regional methodology for Central America to assess tree resources outside the forest.

A mix of satellite remote sensing, aerial photos and ground sampling is used to address the complexity of the resource and to allow dynamic monitoring of resources at the national and regional levels data-gathering on TOF in eight Latin American countries. None of these countries had established a database and the search for information was multisectoral. The statistics on land cover and land use gave some idea of the relative importance of trees outside the forest.

In Kenya, extensive tree planting on farmlands was promoted in the 1970s and 1980s, with land tenure security as a major incentive. There is an increasing trend of tree cover and species diversification on privately owned farms. Assuming that the present rate of increase in tree planting will continue, it was estimated that farms produced about 9.4 million cubic metres of wood in 2000 and will produce about 17.8 million cubic metres in 2020.

Their share of the total wood produced in the medium- and high-potential districts was projected to increase to 80 per cent in 2020. Indeed, while natural stands of trees have declined there has been a corresponding increase in tree planting in much of the densely populated plateaus of Kenya. As natural forests are reduced or become inaccessible, agroforestry systems help people to diversify production and income and to protect themselves from shortages of fuel and wood.

In Bangladesh, natural forest formations cover less than 6 per cent of the country and the population growth rate is extremely high. An inventory of homestead/village forests in the country indicated that trees outside the forest constitute a vital resource for local populations, providing food, fodder and fuelwood. The sampling method was based on dual village/household sampling with an agro-ecological and administrative sampling base. Rural Bangladesh was divided into six major regions considered as agro-ecological strata, each subdivided into *thanas* (administrative entities, subdistricts). The households making up the sampling units were chosen at random from a number of villages.

The inventory sampled data on palm trees and cane as well as trees, bamboo and thickets. The results, expressed per stratum and per inhabitant, provide volumetric data for fuelwood and sawnwood and species data for total amounts under and over 20 cm. This inventory was apparently the first to nationwide assessment of trees outside the classified forests in Bangladesh.

METHODS AND TOOLS FOR FUTURE ASSESSMENTS

A priority challenge of future assessments is to know the state and dynamics of all tree resources both in and outside the forest. A country embarking on a planning exercise cannot confine itself solely to the trees within its forests, especially when its wood resources appear to be insufficient.

The choice of tools and methods used to describe or assess trees outside the forest depends on the scale of analysis, kind of data and degree of exactitude desired. The tools used are not generally specific or new; rather, they are combined and implemented in original ways.

The inventory in Bangladesh described above is one of the numerous examples of methods developed for gathering data on TOF. The Bangladesh study gives evidence of the adaptations needed to bring conventional forest inventory procedures in line with the specific nature of this resource.

In some aspects - e.g. structure, spatial distribution and extent of area cover - trees outside the forest are more difficult to assess than forest formations. The assessment of TOF does not lend itself to the potential cost savings associated with expanded uses of remote sensing technology. Remote sensing by satellite presents more difficulties for assessing TOF resources than for assessing attributes such as forest area. However, satellite data do allow a region to be stratified on the basis of ecological criteria and land cover, providing the basis for a good working document for more specific work in the future. The most commonly used remote sensing technology for TOF resources is aerial photography, which can be used to describe spatial distribution and to distinguish TOF cover classifications, providing the appropriate scale is chosen. However, high costs prohibit widespread use of aerial photography for TOF assessments in most countries. The new 1 m resolution satellite sensors represent a possible future alternative to aerial photography.

Some TOF field inventories are modelled on forest inventory methods and keep to biological and physical criteria; others emphasize social aspects, choosing villages as the sampling units. For measurements on the ground, sampling arrangements designed for forest stands may not be the most effective arrangements for trees. Less traditional sampling plans which would theoretically be better suited to this resource should be tested on various categories of TOF, especially those covering fairly large areas.

Studies of the social and economic benefits or impacts of TOF often rely on household surveys, interviews or standardized appraisals such as rapid or participatory rural appraisal. The integration of the last two approaches - biophysical inventory and socio-economic analysis - is not simple and calls for caution given the great variety of social situations that are only meaningful in the local context. Environmental benefits or impacts of TOF might be indirectly assessed by linking measurable indicators, such as the number and type of trees, with environmental variables such as water quality or erosion. In an urban setting, tree cover might have direct impact on the ambient temperature. Measuring the environmental impact of tree management is an issue for all natural resource planning or management operations.

Assessment of trees outside the forest requires geographical, ecological, biophysical, social and economic data. However, this implies that an important amount of information will have to be carefully processed. The diversity of end-uses for this information, including land use planning and analysis based on inventory, will need to be considered in data assembly and processing and in the presentation of results.

It is important to know the status of trees outside the forest at any given moment, but it is even more essential to be able to trace patterns of change over time in the same area. The two most commonly used approaches have been comparison of aerial photos taken at sufficiently long intervals and surveys among villagers/managers combined with field inventories.

Some countries, such as France and the United Kingdom, have undertaken periodic inventories based on the establishment of permanent plots linked to permanent forest inventories. However, the high cost of this type of operation limits the number of countries able to adopt it. India and Bangladesh are now experimenting with options for the future.

The current trend towards decentralized authority in land use planning suggests the importance of carrying out assessments at the local level, where the geographical, historical and socio-economic context is relatively harmonious.

A minimum number of common rules concerning methods and arrangements is necessary, however, if the data are to be comparable at the country level. Certainly, the technical side of assessing trees outside the forest is complex and more research is needed to better pinpoint the resource.

FOREST: AN AREA OF DENSITY OF TREES

The forest is a complex ecosystem consisting mainly of trees that buffer the earth and support a myriad of life forms. The trees help create a special environment which, in turn, affects the kinds of animals and plants that can exist in the forest. Trees are an important component of the environment. They clean the air, cool it on hot days, conserve heat at night, and act as excellent sound absorbers.

Plants provide a protective canopy that lessens the impact of raindrops on the soil, thereby reducing soil erosion. The layer of leaves that fall around the tree prevents runoff and allows the water to percolate into the soil. Roots help to hold the soil in place. Dead plants decompose to form humus, organic matter that holds the water and provides nutrients to the soil. Plants provide habitat to different types of organisms. Birds build their nests on the branches of trees, animals and birds live in the hollows, insects and other organisms live in various parts of the plant. They produce large quantities of oxygen and take in carbon dioxide. Transpiration from the forests affects the relative humidity and precipitation in a place.

The FAO (Food and Agriculture Organization) has defined forest as land with tree crown cover (or equivalent stocking level) of more than 10% and area of more than 0.5 hectare. The trees should be able to reach a minimum height of 5 m at maturity in situ. Forests are further subdivided into plantations and natural forests. Natural forests are forests composed mainly of indigenous trees not deliberately planted. Plantations are forest stands established by planting or seeding, or both, in the process of afforestation or reforestation.

Forests can develop wherever the average temperature is greater then 10 °C in the warmest month and rainfall exceeds 200 mm annually. In any area having conditions above this range there exists a variety of tree species grouped into a number of forest types that are determined by the specific conditions of the environment there, including the climate, soil, geology, and biotic activity. Forests can be broadly classified into types such as the taiga (consisting of pines, spruce, etc.), the mixed temperate forests (with both coniferous and deciduous trees), the temperate forests, the sub tropical forests, the tropical forests, and the equatorial rainforests. The six major groups of forest in India are moist tropical, dry tropical, montane sub tropical, montane temperate, sub alpine, and alpine. These are subdivided into 16 major types of forests.

India has a long history of traditional conservation and forest management practices. Under British rule, forest management systems were set in place mainly to exploit forests. Nonetheless, there were some attempts to conserve forests and meet the needs of local communities. The Indian National Forest Policy of 1894 provided the impetus to conserve India's forests wealth with the prime objectives of maintaining environmental stability and meeting the basic needs of the fringe forests user-groups. Consequently, forests were classified into four broad categories, namely forests for preservation of environmental stability, forests for providing timber supplies, forests for minor forest produce, and pasture lands. While the first two categories were declared as reserve forests, the rest were designated as protected forests and managed in the interests of the local communities.

Soon after independence, rapid development and progress saw large forest tracts fragmented by roads, canals, and townships. There was an increase in the exploitation of forest wealth. In 1950 the Government of India began the annual festival of tree planting called the Vanamahotsava. Gujarat was the first state to implement it. However, it was only in the 1970s that greater impetus was given to the conservation of India's forests and wildlife. India was one of the first countries in the world to have introduced a social forestry programme to introduce trees in non-forested areas along road sides, canals, and railway lines.

A forest is an area with a high density of trees. There are many definitions of a forest, based on the various criteria. These plant communities presently cover approximately 9.4% of the Earth's surface (or 30% of total land area) and function as habitats for organisms, hydrologic flow modulators, and soil conservers, constituting one of the most important aspects of the Earth's biosphere. Historically, "forest" meant an uncultivated area legally set aside for hunting by feudal nobility, and these hunting forests were not necessarily wooded much if at all.

However, as hunting forests did often include considerable areas of woodland, the word forest eventually came to mean wooded land more generally. A woodland is ecologically distinct from a forest. The latitudes 10° north and south of the Equator are mostly covered in tropical rainforest and the latitudes between 53°N and 67°N with boreal forest.

CHANGES IN FOREST COVER AND CONDITION

An analysis of changes in the cover and condition of the world's forests requires a differentiation between:

1 the increase in forest cover (by afforestation or natural colonization of trees on non-forest land) or decrease (by deforestation) of forest area; and
2 changes in forest condition, either positive (recovery of degraded stands, stand improvement treatments) or negative (decline, defoliation or dieback, effects of forest fires, degradation through

unsustainable exploitation for wood, overgrazing, effects of pests and diseases).

Changes in Forest Cover

Between 1980 and 1995, the extent of the world's forests decreased by some 180 million ha, an area about the size of Indonesia or Mexico. This represents a global annual loss of 12 million ha, an area equivalent to the size of Greece or Bangladesh. During this 15-year period, developing countries lost nearly 200 million ha of natural forests, mostly through clearing for agriculture (shifting cultivation, other forms of subsistence agriculture, the establishment of cash crop plantations such as oil palm, and ranching). This was only very partially compensated for by the establishment of new forest plantations. Over the same period, forests in the developed world expanded slowly (by some 20 million ha) through afforestation and reforestation, including natural regrowth on land abandoned by agriculture.

Forest plantation and natural regeneration on abandoned agricultural land more than compensated for clearing of forests due to urbanization and infrastructure development in most industrialized countries, outside the former USSR. While the gain of forest cover in developed countries was 20 million ha in the period 1980-95, 9 million ha of this increase occurred in the 5 years from 1990 to 1995.

The situation is quite different in the developing world, where deforestation exceeded net afforestation/reforestation, particularly in the tropical zone. As a whole, the annual rate of deforestation in the developing world between 1990 and 1995 was 0.7%, equivalent to 12.6 million ha, with the highest rate in tropical Asia-Oceania, closely correlated with population and income growth. The lowland forest formations were the most affected by deforestation, although the proportional loss during this period was greater in upland formations (1.1% compared with 0.8% in the lowland formations). There is some evidence that the rate of loss of forest cover in developing countries was slowing towards the end of the 15-year period 1980-95. The annual rate of forest loss in the period 1980-90 was 15.5 million ha compared with 13.7 million ha between 1990 and 1995.

Conversion of Forests to other Land Cover

The Forest Resource Assessment 1990 carried out an assessment of the relative importance of the various factors involved in deforestation at regional and global levels between 1980 and 1990 in the whole tropical belt. Among the most significant outputs of the study were the 'area transition matrices' of the type, which indicate transfers from one land cover class to another over the 3068 million ha of the tropical zone covered by the sample. For instance, the first row shows that 1275.9 million ha of the total area of 1368 million ha of closed forest in 1980 remained as closed forest up to 1990 and also shows the fate of the 92.1 million ha converted to other land cover classes (open forest; long fallow; fragmented forest; shrubs and short fallow; other, *i.e.* non-wooded, land cover; and forestry or woody forest and agricultural tree plantations). All these transfers involved changes of woody biomass attempts to capture by replacing each class along the *y*-axis (average biomass per hectare) and showing the importance of transfers along the *x*-axis.

Changes in Forest Condition

Factors affecting the health and vitality of all types of forests have attracted increasing attention in recent years. Outbreaks of fires have made headline news in developed and developing countries alike, while insect and disease attack and generalized declines in forest condition have caused more localized concern. Insects and disease and forest fires exist under 'natural' systems and conditions; what has caused concern has been their spread in artificial systems and in conditions considerably modified by humankind, resulting in direct economic loss.

Fire arising from natural causes (*e.g.* lightning) has been a major influence and is a driving evolutionary force in several forest ecosystems, such as the pine forests of Central America. Fires are important to the health and maintenance of these ecosystems and the total exclusion of fire from them can lead to the build-up of fuel and ultimately abnormally destructive fires that cause the loss of the ecosystem, as happened with the fire exclusion policy of the National Parks Service of the USA and the disastrous fires of 1988.

However, many fires arising from human activities, whether planned or unplanned, have damaged and destroyed large areas of forests and woodlands. Resource managers face a demanding public often with conflicting needs and incomplete information, leading to the development of new policies that take into account public attitudes towards fire management.

At the global level, an agreement was signed at a meeting in 1990 to initiate the International Decade for Natural Disaster Reduction and the target date of 2000 was set for signatory countries to effectively monitor and manage wildfires in their respective countries. In view of the lack of information on causes, extent and the number of fires at present, and the difficulties in collecting data, this target date is unlikely to be met.

Developed Countries

Total forest area is slowly increasing in the developed world, although the need for protection from fire, insects and disease has continued and some aspects of forest condition give cause for concern. Although the widespread decline and death of certain European forests due to air pollution predicted by many in the 1980s did not occur, deteriorating forest condition has remained a serious concern in Europe and North America. The main causes of the decline of forest condition in Europe have been low soil moisture availability due to drought, and high temperatures. These factors, combined with airborne pollution and the use of non-adapted seed sources in forest plantation development, have predisposed forests to attack by insects and disease and to decline. Forest damage due to air pollution is severe in some parts of central and eastern Europe in particular. Air pollution is also said to have a role in the decline observed in certain species in other industrialized countries, such as sugar maple *(Acer saccharum)* in eastern Canada, the high-elevation forests of red spruce *(Picea)* in eastern USA and *Cryptomeria japonica* stands in Japan.

The exclusion of fire may have caused the decline of some forest ecosystems, such as oak stands in central USA or natural eucalypt forests in Australia where fire is an integral part of the ecology of some plant communities and here it may be used to facilitate their natural regeneration.

Despite an overall increase in the number of fires (by some 40% in Europe, 8% in North America and 120% in the former USSR), wildfires over large areas of forest have become less frequent in the industrialized countries. Further, the average area burned by wildfire decreased in most regions between 1983-90 and 1991-94. There has been a slight reduction in total areas burned in the developed world overall as a result of improved prevention, detection and control systems, although there has been an increase of 15% in the former USSR. In 1990, the average area of forest and other wooded land affected annually by fires in Europe, North America and the former USSR together was about 4.26 million ha or 0.22% of their total area of forest and other wooded land. Pests and diseases remain constant threats, particularly to the semi-natural forests and plantations in the industrialized world. Trees are more likely to be attacked when under stress, whether from abnormal weather conditions such as drought or high temperatures, airborne pollution, uncontrolled movement of germplasm or lack of management. Recent outbreaks of pests and diseases that may have been related to stress include the decline of various species of oaks in many parts of Europe and in central Russia, the recurrent infestation of forests in Poland by the nun moth, the attack on beech by the beech scale in western and central Europe, birch and ash diebacks in north-eastern USA, the littleleaf disease of shortleaf pine in southern USA, and the 'x-disease' of pines in southern California.

FOREST CONDITIONS IN DEVELOPING COUNTRIES

Less attention has been paid to forest condition in developing countries because reduction in forest area has attracted more attention. This reduction in density has often been the result of over-exploitation for timber and fuelwood. Overgrazing and repeated bush fires are other significant causes of decrease in density, especially in the dry tropical and non-tropical zones.

Every year, very large areas of savannah woodland and mixed forest-grassland formations are affected by fires set by herders to provide an early flush of green grass when the rains start, particularly in the dry zones of Africa and South America, although no reliable data are available on their extent.

Forests in the humid tropics have also, at times, been affected by large fires, the most serious in recent years having been those associated with the dry weather conditions arising from the El Niño Southern Oscillation (ENSO). In 1997-98, these caused large-scale destruction of logged and secondary forest in Indonesia (particularly Kalimantan, Sumatra and Irian Jaya), Mexico and Central American countries, as well as leading to smoke pollution over wide areas beyond the borders of these countries. The cause of these fires has been burning for land clearance, but it should be noted that the fires are not necessarily in forest; although described as 'forest fires' they are frequently in grassland or on land that has been cleared of forest. It is *estimated* that of the 2 million ha burned in Indonesia in 1997, 150000-200 000 ha may have been in forest. Other serious fires associated with ENSO effects have included that in East Kalimantan, which burned 3.6 million ha in 1983.

Coniferous forests in the humid tropics have often been affected by fires, although it must be noted that fire is frequently necessary to maintain and regenerate them. In the 1980s, the area of pine forest in Honduras and Nicaragua burned annually amounted to some 65 000ha (or about 3.5% of the total pine forest area of these two countries), and widespread fires in natural and artificial tropical pine forests occur in other countries such as Guatemala, Mexico and Indonesia (northern Sumatra).

In the subtemperate and temperate zones of the developing world, fire is also a permanent threat to forests, particularly when they are no longer used by local people for grazing and other purposes. In the 1980s, the average area of forest and other wooded land burned annually was 140 000 ha in the temperate/ subtemperate zones of South America (including southern Brazil). From 1950 to 1990, fires in China are reported to have affected an average of 890 000 ha annually, the most damaging one having been the 'May 6' fire, which burned some 1.85 million ha in the north-eastern province of Heilongjiang in 1987. In the absence of a global statistical fire database, it is difficult to provide an overall estimate of the annual extent of fires in forests and other wooded lands.

A very crude estimate for the temperate/subtemperate and humid tropical zones of the developing world (leaving aside the significant dry tropical zone, for which little reliable information exists) would be of the order of 2 million ha of forest and other wooded land annually during the 1980s. Given the lack of sufficient capacity in fire prevention and control in most developing countries, no significant reduction in wooded areas burned is likely to occur in the near future.

Outbreaks of pests and diseases in developing countries are generally reported for those plantations and planted trees where the impact is most apparent. Introduced pests are usually extremely destructive and several of them have had very damaging effects in recent years in the developing world. Examples of introduced insects affecting plantations and planted trees include the *Leu-caena* psyllid, *Heteropsylla cubana,* which has spread into Asia and the Pacific islands and is now extending across Africa; the cypress aphid, *Cinara cupressi,* which is established in eight eastern and southern African countries and is causing heavy mortality among a number of exotic and indigenous species but especially the important plantation species *Cupressus lusitanica,* threatening its future role in the plantation programmes of these countries; and the European woodwasp, *Sirex noctilio,* which has spread into Argentina, Uruguay and southern Brazil, affecting principally *Pinus taeda* but which may become a threat for the large Chilean plantation estate of *Pinus radiata.*

Strategies to control introduced insects from attacking forest trees include regulatory measures, eradication, integrated pest management, and monitoring of population levels and occurrence of potentially harmful species.

Effective monitoring of insect populations will be especially important in view of the large plantation programmes being carried out to produce industrial roundwood, which often rely on one or a few exotic species. Such programmes may be able to afford the investment that is necessary for control measures. Smaller-scale programmes may have to insure against failure due to unexpected attack by insects or diseases by diversifying into several species adapted to the sites and end-uses.

There is also evidence of forest decline due to a combination of biotic and abiotic factors in the developing world, with air pollution likely to play an increasing role in some cases as industrial and transportation infrastructure develops. Examples of such decline include neem *(Azadirachta indica)* in the Sahel, framiré *(Terminalia ivoren-sis)* in Ivory Coast and Ghana, and *Eucalyptus globulus* plantations in Colombia and Peru, while airborne pollution is affecting forests near large cities and industrial areas in China (*e.g. Pinus massoniana* stands near Nanshan). Despite the overall significance of the corresponding damage and losses, surveys of forest decline and diebacks in developing countries remain all too rare.

RECENT ESTIMATES OF GLOBAL FOREST AREA

The following section is based on 1990 baseline figures, updated to 1995, prepared by the Food and Agriculture Organization (FAO) Forest Resource Assessment Programme, reported in and largely drawn from *State of the World's Forests 1997.*

In 1995 forests were estimated to cover 3454 million ha, or 26.6% of the total land area of the world (Greenland and Antarctic excepted). Almost two-thirds of the world's forests were located in seven countries: Russia, Brazil, Canada, USA, China, Indonesia and Zaire; 29 countries had more than half of their land covered by forest, of which 21 were in the tropical belt. However 49 countries, in addition to the many non-forested small island states and territories, had less than 10% of their land covered by forests. Five entire subregions were in this category: North Africa (1.2% of the land area), Near East (1.9%), Temperate Oceania (6.2%), Non-tropical Southern Africa (6.8%) and West Sahelian Africa (7.5%).

The latest information on the distribution of forest and of forest cover change by ecological zones was provided in 1990; the results of the Forest Resources Assessment 2000 were released by FAO in 2000. In 1990, temperate and boreal forests occupied 1.64 billion ha and tropical forests 1.76 billion ha. A breakdown by ecological zone was available only for tropical forests.The figures on tropical forests show that most (88%) are in lowlands; of these, tropical rain forests accounted for 47% of all lowland tropical forests, followed by moist deciduous forest (38%) and dry and very dry formations (15%).

PATTERN AND PERIODICITY IN INDIVIDUAL FOREST TREE DEVELOPMENT

Tree growth is periodic over short time-spans and follows a definite pattern in the long term. In temperate countries growth is overwhelmingly determined by the season, with growth confined to a period of a few weeks to perhaps several months in any one year while warmth and moisture are adequate. Even in the moist tropics, with year-round favourable growing conditions, trees show periodicity under a measure of genetic control as recorded, for example, from measurements of multinodal tropical pines such as *Pinus caribaea* and fast-growing hardwoods such as *Cordia alliodora.*

As well as periodicity in one year, trees exhibit a strong pattern of growth during their life. In relation to age the pattern is typically sigmoid, whether recording total height, diameter or volume over time.

DYNAMICS OF STAND GROWTH

The dynamic nature of stand growth is a little less obvious than for a single tree, since superficially a stand seems to be a collection of individual trees the relations of which appear constant. That this is not so is seen by considering a small tree in a mature even-aged stand. At regeneration (or planting) this small tree will have been about the same size as its neighbours but subsequently will have competed less successfully for light, nutrients or moisture to become inferior to its neighbours. As a stand develops any apparent uniformity disappears, with some trees developing vigorously and neighbours becoming suppressed, moribund and even dying. Thus the stand is continually changing - it is dynamic. It is not only a collection of individual trees, each with its own genetic potential for using the site, but also a collection of trees that interact and compete with one another.

Seedlings and Regeneration

From the outset trees appear to grow at different rates. Even in new plantations of well-spaced trees of one species, and long before onset of between-tree competition, the growth of individual trees is not identical. Moreover, there is often no correlation between initial postgermina-tion vigour and subsequent growth rate. Small seedlings, or plants from the forest nursery, do not necessarily lead to small slow-growing trees.

Between-tree Competition

As trees grow they eventually begin to compete with their neighbours. This has both an above-ground component, mainly competition for light, and a below-ground element in terms of root competition for nutrients and moisture. Such competition is most readily seen in suppression of side branches on the lower crown and a measurable impact on diameter increment. The timing of the onset of between-tree competition depends on distance between trees, but is usually first measurable in forest stands from about the time of canopy closure. The principal exception is in arid climates where competition between root systems for moisture greatly exceeds that for light, and trees and shrubs remain widely dispersed and rarely in contact above ground.

Differentiation into Crown Classes

As competition begins, growth of some trees slows more than others. These tend to be the smaller trees at the time of canopy closure and the competition reinforces their inferior status. Once dominated, few trees can recover unless a gap develops owing to death of a neighbour. Thus as a stand develops, a range of tree sizes emerges and, traditionally, these are classified into different crown classes according to the tree's relative position in the canopy.

Interventions and Manipulations

These empirical relationships of how trees are observed to grow and how they develop and interact in a stand provide the basis for manipulating their behaviour. Adjusting spacing between trees, both when planted or through thinnings, or influencing the balance of a mix of species in a stand all impact on how a stand develops, on how the increment is distributed and hence on the composition of tree types, trees sizes and the total of woody growth that will result.

In summary, densely stocked forest leads to high volumes per unit area but small mean tree size compared with less well-stocked forest on a similar site and of the same age. In the latter case there will be fewer but larger diameter trees, although total volume of timber may be somewhat reduced. This allows forests to be managed in different ways to yield different assortments of products. How to model these relationships and outcomes in detail, beyond the purely empirical, forms the bulk of this chapter.

Growth Models

The history of forest growth models is not simply characterized by the development of continuously improved models replacing former inferior ones. Instead, different model types with diverse objectives and concepts were developed simultaneously. The objectives and structure of a model reflect the state of the respective research area at its time and document the contemporary approach to forest growth prediction. The history of growth modelling thus also documents the advancement of knowledge in the science of forest growth.

Beginning with yield tables for large regions as a basis for taxation and planning, model development led to regional yield tables and site-specific yield tables and culminated in the construction of growth simulators for the evaluation of stand development under different management schemes. Vanclay (1994) provided an overview about growth and yield management models and their application to mixed tropical forests.

The 1980s brought a new trend with the development of ecophysiological models, which give insight into the complex causal relationships in forest growth and predict growth processes under various ecological conditions. The emphasis in model research has shifted towards ecophysiological models and away from models aimed only at providing growth and yield information for forest management. These models attempt to simulate forest growth on the basis of fundamental ecophysiological processes. The scientific value of ecophysiological models cannot be overrated; however, they will not be applied in forest management for the next few years as they are in many ways not yet sufficiently validated. Also, input and output variables do not yet meet the demand of forest management practice.

A major change has taken place in model conception, *i.e.* the understanding of forest growth on which the model is based. The tables by Weise (1880), Schwappach and Wiedemann resulted from a purely descriptive analysis of sample area data in the form of total and mean values of observed processes of stand development.

These descriptions were later combined with theoretical model concepts that also considered natural growth relationships and causal relations as far as they were known at the time. For example, yield tables for mixed stands of pine and beech created by Bonnemann (1939) characterize growth of beech in the middle and lower storey by mean values. The FOREST model of Ek and Monserud (1974) controls increment behaviour of lower-storey trees by geometrical competition indices, and the ecophysiological growth models of Bossel (1994), Mäkelä and Hari (1986) and Mohren (1987) derive increment behaviour of lower-storey trees from light availability and performance in terms of photosynthesis. The change in model objectives and concepts is closely related to a change in quality of the information generated. Pure management models aim at reliable prediction of forest yield values that are crucial for planning and control in forest management, *e.g.* height and diameter increment and associated economic value.

Ecophysiological models aim at biomass development, nutrient input and loss, etc.; variables relevant to forest management are only of secondary importance in these models. For future planning in modern forestry, models meeting the information demands of ecology as well as of economy will gain in importance.

Ecophysiological models and stand management models can give specific decision support. Ecological and socioeconomic conditions define the framework and thus the 'foundations' for management decisions.

Ecophysiological models can support the ecological elements of the framework, for example the effect of site conditions, species mixture and thinning variants on critical loads, water quality or acidification. Stand treatments of interest can be judged in this way as ecologically acceptable or unacceptable. Management models help to optimize the path from starting point to objective via the given framework, for example they support the decision between different thinning and pruning strategies.

With the shift from tree and stand management models with low resolution to more complex ecophysiological models, different source data are needed for model construction and for the determination of model parameters. Standard datasets derived from research sample plots (diameter, height, etc.) were used for the development of stand growth models for applied forestry. For the construction of single-tree models, additional data are required (crown dimension, tree position, etc.). The transition to ecophysiological models requires an additional database that can only be provided by broadening experimental concepts and cooperation with neighbouring disciplines.

Models are always an abstraction of reality and are greatly influenced by the modeller's knowledge and perception of nature. This applies to the construction of yield tables as well as for eco-physiological models.

Stand Growth Models based on Mean Stand Variables

With a history of over 250 years, yield tables for pure stands may be considered the oldest growth models in forestry science and forest management. They are representations of stand growth within defined rotation periods and are based on a series of measurements of diameter, height, biomass, etc. reaching far back into the past. From the late eighteenth to the middle of the nineteenth century, German scientists such as Paulsen (1795), von Cotta (1821), R. Hartig (1868), Th. Hartig (1847), G.L. Hartig (1795), Heyer (1852), Hundeshagen (1825) and Judeich (1871) created the first generation of yield tables based on a restricted dataset. These original yield tables soon revealed great gaps in scientific knowledge. A series of long-term data-collection campaigns on experimental areas was therefore started. This was the birth of a unique network of long-term experimental plots in Europe that is still under survey.

The second generation of yield tables, initiated towards the end of the nineteenth century and continued into the 1950s, follows uniform construction principles proposed by the Association of Forestry Research Stations (the predecessor organization of the International Union of Forest Research Organizations (IUFRO)), in 1874 and 1888 and has a solid empirical basis. The list of protagonists involved in this work includes Weise (1880), von Guttenberg (1915), Zimmerle (1952), Vanselow (1951), Krenn (1946), Grundner (1913) and, in particular, Schwappach (1893), Wiede-mann (1932) and Schober (1967), who designed yield tables that were conceptually related and is still being used to this day.

A brilliant example of their work is the yield tables for European beech. In the 1930s and 1940s the first models of mixed stands were contructed under the direction of Wiedemann. Data from some 200 experimental areas established by the Prussian Research Station led to the yield tables for even-aged mixed stands of pine and beech, spruce and beech, pine and spruce, and oak and beech. The Second World War prevented Wiedemann from bringing the development of yield tables for uneven-aged pure and mixed stands to an end, but his studies initiated systematic research on mixed stands. Yield tables for mixed stands of this generation were never consistently used in forestry practice as they were restricted to specific site conditions, intermingling patterns and age structures.

Yield tables developed by Gehrhardt (1909, 1923) in the 1920s effected a transition from purely empirical models to models based on theoretical principles and biometric formulae and led to a third generation of yield tables. These models were designed by, among others, Assmann and Franz (1963), Hamilton and Christie (1973, 1974), Vuokila (1966), Schmidt (1971) and Lembcke *et al.* (1975), and at their core is a flexible system of functional equations. These functional equations are based as far as possible on natural growth relationships and are generally parameterized by means of statistical methods. The biometric models are usually transferred into computer programmes and predict expected stand development for different spectra of yield and site classes. A wealth of data were available for the construction of these models and processed with modern statistical methods.

Since the 1960s a fourth generation of yield table models has been created, *i.e.* the stand growth simulations of Franz (1968), Hoyer (1975), Hradetzky

(1972), Bruce *et al.* (1977) and Curtis *et al.* (1981, 1982), which simulate expected stand development under given growth conditions for different stem numbers at stand establishment and for different tending regimes. They describe stand development at different sites and for varying treatments and varying numbers of trees at the time of establishment. Expected stand development under given growth conditions is simulated by means of computer programmes and controlled by systems of suitable functions forming the core of the growth simulator. All information available on forest growth is synthesized into a complex biometric model that simulates stand development for a wide range of possible management alternatives and summarizes the results in tabular form similar to yield tables. These yield tables reflect the stand dynamic for a wide range of imaginable management scenarios. While table and model were identical for the yield tables of earlier generations, simulator-created yield tables now describe just one of many potentially computable stand development courses.

Despite a number of drawbacks, yield tables still form the backbone of sustainable forest management planning. When computing capacities and available data for model construction increased and with the rising demand for information in forestry, mean-value and sum-orientated growth models and yield tables were increasingly replaced by stand-orientated growth models, predicting stem number frequencies, and by single-tree growth models. Prodan (1965, p. 605) commented on the significance of yield tables in the context of silviculture and forest sciences: 'Undoubtedly, yield tables are still the most colossal positive advance achieved in forest science research. The realization that yield tables may no longer be used in the future except for more or less comparative purposes in no way detracts from this achievement'.

Management Models Predicting Stem Number Frequency

With the transition towards new intensive treatment concepts, the demand for information in forestry has changed the emphasis from mean stand values towards single-tree dimensions of selected parts of a stand. This changed demand for information resulted in the 1960s in the creation of the first growth models, which enabled prediction of mean stand values as well as frequencies of single-tree dimensions.

Until then, a stand served as the usual information unit on which all predictions were based; these predictions were now strengthened by statements about stem number frequencies in diameter classes, which are needed for precise prediction of assortment yield and value of a stand. Depending on their concept and construction, stand-orientated growth models predicting stem number frequency are classified into differential equation models, distribution prediction models and stochastic evolution models. Many natural processes in various disciplines of the natural sciences can be described by differential equations. Examples are the differential equations formulating change of yield descriptors for diameter classes of a stand, *i.e.* change of stem number, basal area and growing stock, depending on current yield state values. Stand development then results from the numerical solution of the differential equations. In the 1960s and 1970s, Buckman (1962), Clutter (1963), Leary (1970), Moser (1972, 1974) and Pienaar and Turn-bull (1973) developed stand-orientated growth models based on differential equations.

In the mid 1960s, Clutter and Bennett (1965) proposed a completely new approach to stand growth modelling. They characterized the condition of a tree population by its diameter and height distribution and described stand development by extrapolation of these frequency distributions. The precision of such models is decisively determined by the flexibility of the distribution type on which it is based.

The suitability of different distribution types, for example beta, gamma, lognormal, Weibull or Johnson, has to be assessed individually. Compared with those reviewed earlier, in these models stand development is not controlled by the age function of the individual yield descriptors but by the parameters of the underlying frequency distribution. Models of this type were initially constructed by Clutter and Bennett for North American spruce stands and further developed by McGee and Della-Bianca (1967), Burkhart and Strub (1974), Bailey (1973) and Feduccia *et al.* (1979).

The term 'evolution models' for stochastic growth models is derived from the fact that in these models stand development evolves from an initial frequency distribution, for example from a diameter distribution known from forest inventory. Thus these models, like distribution prediction models, predict frequencies of single-stem dimensions. However, the mechanism accounting for the extrapolation is based on a Markov process, giving the transition probability for the shift between the diameter classes. Stochastic growth models were introduced to forestry science with the pioneering investigations by Suzuki, and they continue to be linked to his name today. His growth models, for Japanese *Chamaecyparis* pure stands for example, have been consistently developed by Sloboda (1976) and his team since the mid 1970s; they are mainly interested in adapting the models, which are orientated to Japanese conditions, to the issues of German forestry and in model validation based on permanent test plot data. Stand-orientated growth models based on stochastic processes have been developed by Bruner and Moser (1973) and Stephens and Waggoner (1970) also for mixed stands.

Single-tree Orientated Management Models

Single-tree models describe the stand as a mosaic of single trees and model individual growth and interactions with or without consideration of tree position. This has paved the way for the design of models of pure and mixed stands of all age structures and intermingling patterns. An equation system that controls growth behaviour of single trees depending on their constellation within the stand is the central module of all single-tree models. Position-independent or position-dependent competition indices are used to quantify the spatial growth constellation of each tree and to predict its increment of height, diameter, etc. in the following period. Compared with stand-orientated growth models based on mean stand descriptors and those predicting stem number frequencies, single-tree models work on higher resolution. The information unit in singletree models is the individual tree. However, results of lower-resolution models, for example mean tree development or diameter frequency distributions, can also be derived from single-tree model results by integration. Information about stand growth then results from summarizing and aggregating each individual single-tree development for a given growth period. Recent single-tree models are programmed to enable the user to influence a simulation run interactively. This allows stand development to be followed step by step during the simulation and permits the user to specify other factors (*e.g.* thinning or influence of disturbance) at any time during the simulation process, thus influencing or diverting the current course of stand development.

After parameters for the control of the singletree model have been set, tree characteristics at the beginning of the prediction phase for the test area to be investigated are fed into the computer as initial values for the simulation. This tree list can contain data on tree species, stem dimensions, crown morphology, stem position and other data about the stand individuals. These data usually originate from single-tree-based inventories of indicator plots. Starting with these initial values, change (*e.g.* mortality or development of diameter, height or crowns) for all stand members depending on individual growth conditions is predicted using an appropriate control function; this is done for a first growth period, for example 5 years. Once the tree list has been processed, change of growth conditions (*e.g.* due to thinning or disturbance) can be specified prior to continuing to the next increment period. This will now influence single-tree growth in the following period. The modified state values of all trees resulting at the end of the first growth period also represent the initial values for the second growth period.

These values are repeatedly extrapolated in every simulation cycle and interim results are given. The simulation continues until the envisaged prediction period has been completed step by step. In most models, time steps are 5 years, sometimes only 1 or 2 years. By removing single trees during a simulation run, the growth constellation and growth behaviour of the remaining individuals change in the next growth period. Growth reaction of the stand is thus explained by the reactions of all single trees to this intervention. By relating stand development back to growth behaviour of single trees and by modelling single-tree dynamics depending on growth constellation within the stand, single-tree models, after being initialized accordingly, enable evaluation of a wide range of treatment programmes.

The first single-tree model was developed for pure Douglas fir stands by Newnham (1964). It was followed by the development of models for pure stands by Arney (1972), Bella (1970) and Mitchell (1969, 1975) and colleagues. In the mid 1970s, Ek and Monserud applied the construction principles for single-tree orientated growth models for pure stands to uneven-aged pure and mixed stands (Ek & Monserud 1974; Monserud 1975). Munro (1974) distinguished distance-dependent and distance-independent single-tree models, the former being able to refer to data about stem position and stem distance for the control of single-tree growth.

The worldwide bibliography of single-tree growth models compiled by Ek and Dudek (1980) lists more than 40 different single-tree models, which are grouped into 20 distance-dependent and 20 distance-independent models. Single-tree models developed since the 1980s in many ways go back to the methodological bases of their predecessors; however, owing to the rapidly improving technology of modern computers they are far more user-friendly than older single-tree models.

Ecophysiological Growth Models

All the models mentioned above rely on growth and yield data from long-term observation plots and hence have the advantage of being validated empirically. However, there is a drawback to historically deduced data in as much as growth conditions undergo changes, and reaction patterns from the past cannot simply be projected into the future. In the 1970s, model research was pointed in a new direction with the creation of high-resolution ecophysiological process models, which account for metabolism, organ formation, assimilation and respiration as well as biochemical and soil chemistry reactions. Pioneers of the ecophysiological process model for forest stands are Bossel (1994), Mäkelä & Hari (1986) and Mohren (1987). The term 'process model' is slightly misleading in the sense that all forest growth models describe processes. Only the temporal and spatial scales of modelled processes become more detailed and accurate in the transition from yield table models via single-tree management models and succession models to growth models based on ecophysiological data.

The development of modern process models begins with a systems analysis and the selection of characteristic system components. A system to be analysed and modelled is first described using methods of systems analysis. Results of this description can be transferred into a system diagram. The description breaks the system down into system components characteristic for all biological systems and identified by different symbols in the system diagram. By system parameters we mean those that remain constant during the lifetime of the system. Exogenous parameters are variables that control the system but which cannot be influenced by the system, *e.g.* stress caused by air pollutants. State variables are the actual output value of the model; their current values reflect the system's state. Important state variables in stand models are accumulated carbon quantities in needles, branches, stem and roots. The initial values of the state variables give the starting values of a system and thus crucially influence its further development. In a growth model for example, stem number and initial stand structure have to be specified as initial values. The rate of change of the state variables controls change, *i.e.* input and output of state variables.

Examples are mortality rates or respiration rates, which control the change of the carbon quantities accumulated in the different components. Intermediary variables change simultaneously with the state variables and feed back into the system. The system components are indicated in the system diagram with different symbols and their interrelations are identified by arrows.

The model thus outlined is transferred into a mathematical model and subsequently into a computer programme. For this, the system components and links are described by mathematical or logical relationships. Once the complete model is constructed, the causal relations implemented are parameterized. The system behaviour can be simulated with the developed computer programmes. All suitable information known about the system is therefore consolidated in the system components and the system structure. The process of system analysis and model development concludes in the validation of the final model. For validation, *i.e.* testing if the causal relations assumed in the model realistically reflect growth of stands or single trees, empirical yield data can be used. If necessary, individual model assumptions are corrected or model parts revised.

A vastly improved understanding of ecophysiological processes in forest ecosystems paved the way for this model approach and it was the actual modelling of these processes that provided an idea of the functioning of the overall system. A further impetus to process model development was the need to understand and predict the reactions of forest ecosystems to an increasing number of adverse effects, such as industrial emissions, rise in atmospheric CO_2 and climate change. In the context of environmental instability, high-resolution and accurately detailed process models are certainly the ideal approach for understanding and predicting forest ecosystem behaviour. However, there are particular constraints in developing and applying process models due to considerable gaps in our knowledge of part processes in assimilation organs and in the soil.

Also, the scaling-up of part processes to the behaviour of the overall system is still largely unresolved. Moreover, the introduction of process models still requires intensive research and extremely high-powered computers that are only rarely available in practice. To date, process models are therefore primarily research instruments rather than forest management planning tools.

Gap Models and Biome Shift Models

In the view of modern theoretical ecology, a spatially extensive system is composed of mosaiclike subunits and can be studied by analysing these subunits. Watt (1925, 1947), Bormann and Likens (1979) and others transferred this view of extensive ecosystems to the study and model representation of the growth dynamics of pure and mixed stands. This laid the foundations for the concept of gap models suitable to predict succession. According to this concept, a forest stand is an aggregation of gaps. The size of these gaps corresponds to the extent of a potential crown area of a dominant tree or tree group (areas of 0.04-0.08 ha).

The actual information unit is the tree group in the gap; stand development results as the sum of the total spectrum of contributing gaps. Gap models imply that forest development in a gap occurs in a fixed cycle: A gap results from exploitation or death of a dominant tree, and thus the growth conditions of understorey trees improve and natural regeneration occurs. Growing trees successively close the gap and a new overstorey develops. The cycle is repeated with further losses of dominant trees. Growth models using this approach were predominantly employed for investigations of competition and succession in semi-natural stands.

Gap models, such as those designed by Shugart (1984), Pastor and Post (1985), Aber and Melillo (1982) and Leemans and Prentice (1989), are primarily aimed at mixed stands. While in the models described above increment-determining factors have effects on stands or individuals respectively, gap models describe tree growth that depends on growth conditions in the individual gap. Gap models simulate growth dynamics for single trees or tree classes in a gap; it is therefore possible to generate information about the development of diameter, height and volume of single trees as well as stands.

However, regarding input and output variables they are less dependent on information available from, or required by, forestry practice; rather, they aim at predicting long-term succession in natural forest stands and the effects of altered growth conditions. The FORMIX2 model for virgin and logged Malaysian lowland dipterocarp forests is an example of an ecophysiological-based gap model with output variables that is useful as decision support in forest management.

Biome shift models, such as those of Box and Meentemeyer (1991) and Prentice *et al.* (1992), establish statistical relationships between regional climate and vegetation type. Based on relevant climatic conditions, the nature of potential biomes, *i.e.* communities, may be predicted on a regional and even global scale. Of all the models under discussion, these are the ones that provide the highest aggregation of data on vegetation development and forest growth. They have therefore gained increasing importance in research on global change.

Hybrid Models for Forest Management

The transfer of specific components of ecophysio-logical models (based on solid process knowledge) into stand or single-tree management models (based on long-term experimental plots and increment series) leads to what Kimmins called 'hybrid growth models'. Models of this type were constructed by, among others, Botkin *et al.* (1972) and Kimmins (1993). Their objective is to make the best possible use of the newly acquired knowledge of ecophysiological processes combined with historical increment observations to assist in forest planning and management. On account of the implemented relationship between site conditions and species-specific growth, they can be used for pure and mixed stands. In the past 100 years mixed stands have gradually become the focus of forest research, particularly on account of studies by Gayer (1886), Wiedemann (1939b) and Assmann (1961), but to this day growth models for mixed stands are scarcely used as quantitative planning tools.

Only very recently have models created by Kolström (1993), Nagel (1996), Pretzsch (1992), Pukkala (1987) and Sterba *et al.* (1995) found use in forestry practice for planning work in pure and mixed stands. These are in effect site-sensitive single-tree models constructed from a broad base of ecophysiological and growth and yield data. Version 2.2 of the SILVA model, developed in Germany for pure and mixed stands, belongs to the category of hybrid models and may be used as an example to explain the functional principles underlying this approach.

Management Model SILVA 2.2 for Pure and Mixed Stands

SILVA reflects the spatial and dynamic character of mixed-stand systems in as much as it models spatial stand structures at 5-year intervals. This permits the recording of the individual growth constellation of every tree and the control of tree increment in relation to growth constellation and the original dimensions of the tree. The external variables determining tree increment and stand structure are treatment, risk and site factors. The model simulates the effects that tending, thinning, regeneration and natural hazards such as storms and wind have on the stand dynamic.

The feedback loop, stand structure —> tree growth —> state of tree —> stand structure, forms the backbone of the model. The step-by-step modelling of the growth of all individual trees via differential equation systems provides information about the development of assortment yield, financial yield, stand structure, stability and diversity of the stand over and above the data, required in yield calculations, on height, diameter at breast height, number of stems, etc. Input and output data used in the model correspond to the data available from, or required in, forestry practice, for example only site variables available on a large scale are considered. With models of this type a weighting between yield-related, socioeconomic and ecological aspects of stand development in pure and mixed stands becomes possible. Parametrization relies on yield and site characteristics of pure and mixed stands that have been under observation for over 100 years.

The position-dependent individual tree model SILVA 2.2 breaks down forest stands into a mosaic of individual trees and reproduces their interactions as a space-time system. It can therefore be used for pure and mixed stands of all age combinations. Primarily it is designed to assist in the decision-making processes in forest management. Based on scenario calculations SILVA 2.2 is able to predict the effects of site conditions, silvicultural treatment and stand structure on stand development, and therefore also serves as a research instrument.

A first model element reflects the relationship between site conditions and growth potential and aims at adapting the increment functions in the model to actual observed site conditions. With the aid of nine site factors reflecting nutritional, water and temperature conditions, the parameters of the growth functions are determined in a two-stage process. The stand structure generator STRUGEN facilitates the large-scale use for position-dependent individual tree growth models. The generator converts verbal characterizations as commonly used in forestry practice (*e.g.* mixture in small clusters, single tree mixture, row mixture) into a concrete initial stand structure with which the growth model can subsequently commence its forecasting run.

The three-dimensional structure module uses tree attributes such as stem position, tree height, diameter, crown length, crown diameter and species-related crown form to construct a spatial model of the stand in question. The thinning model is also based on individual trees and can model a wide spectrum of treatment programmes. The core of the thinning model is a fuzzy logic controller. In the simulation studies described below the thinning model simulates various thinning methods (thinning from below and selective thinning) and thinning intensities (slight, moderate and heavy). The competition model employs the light-cone method and calculates a competition index for every tree on the basis of the three-dimensional stand model. The allocation model controls the development of individual stand elements. Tree diameter at height 1.3 m, tree height, crown diameter, height of crown base, crown shape and survival status are controlled, at 5-year intervals, in relation to site conditions and interspecific and intraspecific competition. Finally classical yield information on stand and single-tree level for the prognosis period are compiled in listings and graphs. Additional information on stem quality, assortment and financial yield complete the growth and yield characteristic. At every stage of the simulation run, a programme routine for structural analysis calculates a vector of structural indices that serve as indicators for habitat and species diversity and form a link to the ecological assessment of forest stands.

The algorithmic sequence for predicting forest development comprises the following steps. Step 1 is the input of data on the initial structure and site conditions of the monitored stand. In step 2, the parameters of the growth functions are adapted to actual site conditions.

Once the starting values for the prognostic run are complete, monitoring can begin. If there are no initial values, for example stem positions are unknown, the missing data can be realistically complemented with the help of the stand structure generator (step 3).

Once the spatial model has been constructed (step 4) the silvicul-tural treatment programme is specified in step 5. The competition index calculated for each tree through the three-dimensional model in step 6 is used, in step 7, to control individual tree development. Steps 4-7 are repeated until the entire prognostication period has been run through in 5-year steps.

To date, model research has had little success in substituting the yield tables for pure stands by an improved information system for pure and mixed stands. This can in no way be attributed to a deficit in methodological principles, data or technical equipment. Rather, the causes lie in the fact that new models are not properly adapted to practical requirements. The recent introduction of the growth model SILVA 2.2 for forest management use led to a range of operational requirements and outputs demanded from the management models that will be used in decision-making processes at stand and forest enterprise levels.

1. The natural management of forests is currently making great headway. In the long run only those growth models capable of simulating the growth of pure and mixed stands of all age compositions and structural patterns will find approval.
2. Models need to be operable at stand and forest enterprise levels and able to simulate growth behaviour under different thinning regimes and different processes of artificial and natural regeneration.
3. Flexibility of the model is essential so as to permit simulation of growth reactions to site alterations and interference factors on a large regional scale.
4. Apart from tree and stand characteristics such as volume production, assortment yield, wood quality and financial yield should also include structural parameters determining the recreational and protective functions of forests as well as indicators showing the impact of hazards or ecological instability.
5. Forestry practice is interested, first and foremost, in calculating scenarios at stand and forest enterprise levels. This can only be achieved if input and output data of the model consider what

information is available and which data are needed in forestry practice. Furthermore, achieving this goal also depends on whether the model forms part of a comprehensive forestry information system and, lastly, whether hardware specifications are acceptable in practice.

For decades forestry practice has been hoping for improved growth models to assist with research, planning, operations and control in forest management. The general acceptance of new models by practitioners calls for close cooperation between forest science and forest practice, from the design and development of the model to its actual introduction in forest management.

Likewise the only capital equipment required is a weaving needle, which is a very cheap item to buy. Similarly, wood is often used to make a variety of household implements and agricultural tools, or is carved to make products aimed at the tourist market. Once again, interviews with resource makers stress that the skills used to make these objects can be acquired in a very short time and the capital equipment required is minimal. Finally, where NTFP foods are processed to make commercial products, such as fruit wines, palm wines, dried meats, dried fish and dried insects, again the skills and capital equipment required are low: it is essentially a matter of allocating sufficient household labour to complete the task. Due to the difficulties that rural households face in capital accumulation, because formal credit markets generally fail in rural areas and because education levels are low, rural households are usually endowed with unskilled labour but poorly endowed with capital. So another reason for the extensive use of NTFPs by rural households is that the factor inputs required to collect and process NTFPs (unskilled labour, little capital equipment) match closely the factor endowments of the classic rural household. (These required factor inputs, combined with open access to the NTFP resource, means that NTFP activities have very low entry barriers. In such circumstances we would therefore expect the returns to NTFP activities to be correspondingly low.)

Fourth, households also use NTFPs in response to the general riskiness of rural economic activities. The use of NTFPs, particularly what have been classified as 'minor forest products', during times of household stress is an observation common to much of the NTFP case study literature. NTFPs display a certain degree of non-covariance with respect to agricultural output. When crops fail due to drought or disease, or when shocks hit the household such as unemployment, death or disease, some NTFPs will still be available for the household to either consume or use to generate cash income to purchase its essential needs. Thus it is economically rational for risk-averse rural households to hold a portfolio of production opportunities that exploits this non-covariance. In this sense, commonly held forests and woodlands can be regarded as providing a set of back-stop resources that insure households against the failure of other, higher return production activities. Thus, even if the returns per hectare of woodland were systematically below that of agricultural land, rural villages would still retain common lands for this purpose.

Finally, the argument comes full circle. One of the reasons that rural households use NTFPs so extensively is precisely because they are open-access resources. There are a number of reasons for the survival of communally held resources, for example the insurance element; however, a major reason for the survival of such resources is the very low income levels of rural households. Resource privatization is an expensive business: households taking charge of privatized resources usually need to spend time and money on creating exclusion (*e.g.* by building fences), on enforcement (*e.g.* employing guards to monitor incursions) and on punishing infractions (by going through the courts). Particularly where a resource has previously been held in common, privatization is often strongly contested and consequently these various costs can be expected to be high. Where households are generally poor, the private costs of resource privatization are likely to be far higher than the potential private gains from control of the resource. Thus the low incomes of rural areas underpin the existence of commonly held resources, which in turn underpins the extensive use of NTFPs by those very same low-income households (on the relationship between efficient property rights regimes and transactions costs.

FOREST PLANTATIONS

Of particular interest in many arid regions of the world are forest plantations that are established as windbreaks and shelterbelts, for sand dune stabilization, canal side and riverside plantations, and as amenity plantations.

WINDBREAKS AND SHELTERBELTS

In arid zones, the harsh conditions of climate and the shortage of water are intensified by the strong winds. Living conditions and agricultural production can often be improved by planting trees and shrubs in protective windbreaks and shelterbelts which reduce wind velocity and provide shade. Windbreaks and shelterbelts, which are considered synonymous in this manual, are barriers of trees or shrubs that are planted to reduce wind velocities and, as a result, reduce evapotranspiration and prevent wind erosion; they frequently provide direct benefits to agricultural crops, resulting in higher yields, and provide shelter to livestock, grazing lands, and farms. A main objective of windbreaks and shelterbelts is to protect the agricultural crops from physical damage by wind.

Other benefits include:

- Preventing, or at least reducing, wind erosion;
- Reducing evaporation from the soil;
- Reducing transpiration from plants;
- Moderating extreme temperatures.

Quite often, protection can be combined with production by choosing tree and shrub species that, apart from furnishing the desired sheltering effect, yield needed wood products.

DESIGN OF WINDBREAKS AND SHELTERBELTS

When considering windbreak or shelterbelt planting, three zones can be recognized: the windward zone (from which the wind blows); the leeward zone (on the side where the wind passes); and the protected zone (that in which the effect of the windbreak or shelterbelt is felt).

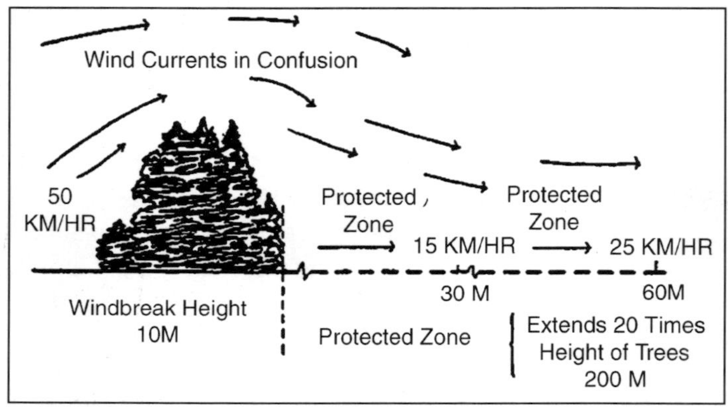

Fig. Functioning of a Windbreak

The effectiveness of the windbreak or shelterbelt is influenced by its permeability. If it is dense, like a solid wall, the airflow will pass over the top of it and cause turbulence on the leeward side due to the lower pressure on that side; this gives a comparatively limited zone of effective shelter on the leeward side compared to the zone that a moderately permeable shelter creates. Optimum permeability is 40 to 50 per cent of open space, corresponding to a density of 50 to 60 per cent in vegetation. Gaps in the barriers should be avoided. Permeability of dense shelterbelt can be improved by pruning lower branches at 0.50-0.8 m from the soil level.

It is generally accepted that a windbreak or shelterbelt protects an area over a distance up to its own height on the windward side and up to 20 times its height on the leeward side, depending on the strength of the wind. In reducing wind speeds, narrow barriers can be as effective as wide ones. Furthermore, a narrow shelterbelt has the advantage of occupying less land.

The shape of the cross-section of a windbreak or shelterbelt determines, to a great extent, the sheltering effect. To a large extent, the choice of tree or shrub species to plant, along with their planting arrangement, dictates the cross-sectional shape. In general, an inclined slope facing the wind should be avoided, as it only deflects the windflow upward. Barriers with a clear vertical side provide best wingspread reduction. When designing a windbreak or shelterbelt, the direction of the wind must be considered. A barrier should be established perpendicular to the direction of the prevailing wind for maximum effect. To protect large areas, a number of separate barriers can be created as parts of an overall system. When the prevailing winds are mainly in one direction, a series of parallel shelterbelts perpendicular to that direction should be established; a checkerboard pattern is required when the winds originate from different directions. Before establishing windbreaks or shelterbelts, it is important to make a thorough study of the local winds and to plot on a map the direction and strength of the winds.

Selection of Tree and Shrub Species

In the selection of tree or shrub species for windbreaks or shelterbelts, the following characteristics should be sought:

- Rapid growth;
- Straight stems;
- Wind firmness;
- Good crown formation;
- Deep root system, which does not spread into nearby fields;
- Resistance to drought;
- Desired phonological characteristics (leaves all year long or only part of the year).

Planting Techniques

Planting techniques for windbreaks and shelterbelts are identical to those in other tree and shrub planting programmes. However, as windbreaks and shelterbelts require a high plant survival rate, as well as uniform and rapid growth, supplementary irrigation may be required during the establishment phase. Gaps cannot be tolerated and, when plants are lost, replacement must be prompt. Although in theory, one-row barriers should suffice, experience has shown that the most effective windbreaks and shelterbelts are those consisting of several rows of trees. Quite often, initial spacing is 3 meters between the rows, with trees 2 meters apart in the rows. Where trees or shrubs have long roots that could extend into agricultural fields, vertical root pruning may be recommended; this can be done with special equipment or by digging trenches. A triangular arrangement of plants is frequently prescribed.

Management Practices

Once established, the effectiveness and longevity of a windbreak or shelterbelt depends on its maintenance. As the trees and shrubs mature, they change in shape and appearance, which necessitates some level of maintenance to ensure a continuing shelter effect. Pruning may be required to stimulate height growth, while thinning can boost diameter growth. To keep a barrier at the desired density and permeability, occasional pruning or removal of plants may be necessary. If trees or shrubs are damaged by wind or pest attacks, a control is also needed. In all of these cases, the management practices depend on the desired composition of the barrier and the species used. Since these management practices can involve the removal of woody parts, the use of tree or shrub species that make fuelwood or fodder available on a continuous basis is desirable.

A windbreak or shelterbelt has a life that is dependent on the trees or shrubs of which it is composed. Therefore, to be able to furnish permanent shelter, a renewal plan should be adopted. To renew a barrier consisting of many rows, felling the rows on the leeward side and then replanting them is often recommended. If the windbreak or shelterbelt consists of one row, a new row may be planted parallel to the the old one; when the new row has matured, the old one is removed. To renew narrow windbreaks or shelterbelts arranged into a system, new belts can be planted midway between the existing barriers which, in turn, are to be removed when the new ones become effective.

When windbreaks or shelterbelts are established on grasslands or other areas where animals are allowed to graze, special attention must be paid to the protection of the barrier; this can be done by planting thorny vegetation or by using a barbed-wire fence along the edges of the barrier.

Sand Dune Stabilization

Sand dunes result from wind erosion. They are formed in many arid lands when winds regularly blow over poorly-vegetated areas. Sand dunes that are not covered with vegetation (because of overcropping or overgrazing) move in the direction of the wind at a speed which can approach 10 meters a year, endangering agricultural crops, forest plantations, irrigation canals, and roads. To prevent this encroachment, the sand dunes must be stabilized; one method of sand dune stabilization is to establish a vegetative cover.

In general, two types of sand dunes are recognized: coastal dunes and inland dunes. Techniques of stabilizing these two types of sand dunes through the establishment of a vegetative cover.

Stabilization of Coastal Dunes

Coastal dunes originate from sand thrown up onto the shore by waves. At low tide, the sand dries and is blown away by the wind. When protective vegetation beyond the beaches is destroyed, coastal dunes move inland. To stop the advancement of coastal dunes, an artificial foredune should be constructed about 50 meters from the floodline. Normally, this initial barrier is built one year before a planting programme begins.

One method of building a foredune is by mechanical fixation of the sand by fences or palisades, 0.5 to 1 meter high. The materials used for the fences or palisades may include twigs from trees or shrubs, brushwood, grass sheaves, reeds, bushes, palm leaves, old railroad ties, used oil drums, and earth. When the prevailing wind has a prevailing direction, parallel lines of palisading are sufficient; however, a checkerboard system is advisable where fluctuating winds are common. Sand piles up behind the palisade and, when the artificial dune that is formed reaches a height of 0.5 to 0.75 meter, a second palisade is built on top of it. Sometimes, the original barriers can be raised, when necessary, instead of building a new palisade.

Once the foredune is established, it is possible to stabilize the sand behind it by seeding or planting a vegetative species that provides good ground cover and is able to withstand (at least partially) covering by sand.

Sand dune fixation also can be done by mechanical mulching; that is, the spreading of solid material on the surface of the sand. Chemical fixation can also be employed. Chemical fixation consists of stabilizing the sand surface by covering it with a continuous crust of sprayed chemical substances, such as petroleum derivatives or latex mixtures. Vegetative establishment is usually a follow-up or a concurrent operation. Chemical fixation is advisable when the cost of labour is high and the chemicals are readily available.

Sand dune stabilization with plant species is more permanent than mechanical mulching and chemical fixation techniques which are, in most cases, only temporary measures.

Stabilization of Inland Dunes

Inland dunes originate from sand produced by the weathering of rocks, mainly sandstone. The fine fraction can be blown far away, while the heavier fraction is blown short distances and forms dunes. Such dunes can pose serious stabilization problems, especially when the dunes are large and active. One way to combat this problem is by creating an artificial dune at the windward end of the dune. The method followed is similar to the one used to create the foredune in the stabilization of coastal sand dunes. The stabilization of inland dunes also follows the same general lines.

When a specific area of value (for example, an oasis) is threatened, protective work is initiated as close as possible to the area of concern, with the work gradually progressing towards the sand source area.

Planting Techniques

Planting of vegetation is the best and most permanent method of coastal and inland sand dune stabilization; both direct and indirect benefits can be realised, including:

- Protection (of roads, canals, agricultural lands, and industrial areas);
- Wood production (fuel, lumber, etc.);
- Protection of watershed areas and water supplies;
- Livestock grazing benefits (including fodder);
- Wildlife benefits, recreation, and other amenities;
- Public works to combat unemployment.

The choice of vegetative species for planting should be based on studies of the natural vegetation in the area and on the environmental conditions. As planting of vegetation on sand dunes frequently consists of afforestation practices, it is recommended that species trials be included in the planting programmes to permit an evaluation of tree and shrub species for long-term use.

In practice, it is often necessary to plant relatively large containerized plants close together (1 × 1 meter) on the windward side, but they can be planted further apart (2 × 2 meters) on the sheltered side. Irrigation for initial establishment may be required to help the plants survive until they have sufficiently deep root systems. If water is not available in adequate quantities for irrigation to take place on a long term basis, it is advisable to irrigate (at least) during the first two or three months after planting, at weekly intervals.

Concerning maintenance, hand weeding is preferred to avoid problems of machinery traction in the sand. As a rule, all livestock movement and other traffic should be eliminated on the sand dunes; when necessary, delimited and protected passages for livestock can be established.

CANAL–SIDE PLANTATION

In may arid countries, wherever rivers are available, efforts have been made to utilize the water for irrigation purposes through the construction of dams or using lift irrigation for the agricultural needs. Several thousands of kilometers of irrigation canals have been laid. The banks of such canals are available for planting purposes and constitute a considerable area for production of timber and firewood for the rural population. Full advantage is being taken of this in many countries like China, Egypt, India and Pakistan. A few rows of trees, varying from 4 to 6, are generally planted on each bank of the canal with an espacement depending on the characteristics of the species and the type of produce desired.

When designing a canal plantation, the requirement may be the same as for the design of irrigated plantations with respect to climatic and soil conditions and to supply and quality of water. However, it should be remembered that the only water supply available to the trees is seepage from the canal into the root zone. In some places, it is cheaper to grow trees and thus utilize the seepage water rather than prevent seepage by canal linings of concrete, asphalt or other material.

Choice of species for canal side plantations should take into account both the particular character of the plantation and its purpose. The roots of the trees should strengthen the banks of the canal and the trees should keep the canal and its banks well shaded in order to suppress weed growth and reduce evaporation. Species that tend to increase water seepage through the sides and bottom of the canal should be avoided. Where canals have an intermittent flow, such as flood discharge canals, only trees able to adjust to varying water levels in the soil can be used. Species that reproduce by suckers such as Robinia pseudo acacia should not be planted along canals. Plantation techniques should favour deep planting and roots should be planted in the moist layer.

RIVER-BANK PLANTATIONS

There are many areas where river lengths are considerable. The ground on either side of the river is partly within the reach of the high level of water during the period the rivers are in flood. Beyond this level—and on the fringes of the agricultural land, strip plantation can be established to produce wood, fuelwood and fodder. Generally, the width of such strips is limited but does constitute a useful and productive linear plantation. Underground water is available at different levels. The species to be planted should be matched with this water level variation.

Spacing within and between the rows depends on the characteristics of the species and the rotation planned for the crop. In the more arid areas, trees with xerophytic habit constitute the outermost rows while those close to the river bank are the ones with higher water requirement. In such locations, phreatophyte species such as Populus spp., Acacia nilotica, Dalbergio sisso, Prosopis spp. can be planted.

AMENITY PLANTATIONS

This type of plantation includes trees planted in gardens and parks, street planting, green belts around villages and cities, trees planted along roadsides to reduce noise and beautify the homestead or landscape. In arid zones "beautifying the landscape" usually means changing the countryside from its normal brown colour to green, or "greening" of the landscape.

Tree Planting in Gardens

Tree planting in gardens is usually regarded as a beautification of the home environment but it also has a profound effect on man's psychological attitude towards life. A house located in a barren landscape without trees and shrubs lacks appeal and there is certainly a different psychological attitude towards it, compared with a home which has been beautified and protected by wisely selected trees and shrubs. Tree planting in gardens also enhances self-esteem. The gardener identifies with his garden and builds a personal relationship with it.

The garden becomes an extension of himself, a visible representation of his individuality. When it blooms, he has evidence of his success. He also becomes aware that a number of people he does not know pass by each day and enjoy his garden. He has given them an anonymous gift. All of this enhances his self image, helping to create self-esteem. The gardener, feeling better about himself, feels better about where he lives.

In selecting species for tree planting in gardens, the following should be considered:

- Trees have several functions to play in plantations around houses. The first point to be considered is therefore to decide on the purpose for which planting is to be done, that is: shelter, shade, ornament, hedges, scent and odours, as source of fruits or nesting sites for birds. Some plants will fill only one or two of these objectives while others may fill more.
- Trees selected should be suitable to local climate and soil conditions.
- Adaptability to pruning. The degree to which trees and shrubs will tolerate pruning and pollarding is an important consideration for planting around the homestead.
- Evergreen or deciduous habit: most arid-zone species are evergreen but many deciduous trees are also used for planting in gardens. In the Mediterranean arid region, deciduous trees have certain advantages and disadvantages when compared with evergreens. The most important advantage is their ability to provide shelter in summer but allow sunlight to penetrate in winter. The most important disadvantage is that the leaves need to be gathered during and after leaf fall.

Planting in Parks

The first objective for planting in parks in arid zones is for shelter from the sun and dust. Where trees are not growing naturally, shelter may be provided by planting trees in favourable sites. A second objective is to enhance the beauty of the park by planting trees of different colour, shape and size. Planting formal layout patterns of trees in long straight rows does not usually fit into the landscape pattern, and planting in small clumps or as individual dispersed trees is usually more suitable. Choice of species should not be dictated by the value of trees for wood, nor planting expenses weighed against the value of the wood produced. Costs for planting in parks for recreation should take into account aesthetic benefits and the intangible gains in the health and well-being of the people.

Street Planting

Street plantings are often the responsibility of local municipalities and are made to beautify the cities, provide shade and control outdoor noise and traffic pollution. In recent years, many of the towns and cities in the world have learned through experience that paving roadways, streets, and sidewalks does not complete the job. In most communities where there has been an increase in population and greater congestion of motorized traffic, it has become important that the city municipality assumes its responsibility of providing the amenities which add much pleasure to life in the city.

For use in street planting, trees should:

- Be easy to establish, preferably with the ability to be transplanted as advanced nursery stock, and grow relatively quickly to the stage that they provide some amenity value;
- Be healthy in the environment, relatively long-lived and not subject to wind-throw or the breakage of large limbs;
- Be as maintenance-free as possible. Trees requiring permanent pruning and removal of fallen leaves will have a high maintenance cost.
- The form and height of the species must be suitable for the width of the street in which they are to be planted;
- Whether to use a single variety or mix on the same street is a matter of taste. A variety might be used depending on the geographic section of the country and the width of the street.

Greenbelt Planting

Several cities in arid zones have established in their municipal area green belts with a number of purposes:

- To enhance the beauty of the site;
- To provide a recreation area for the urban dwellers;
- To reduce the harmful effect of dry winds and dust storm and control sand encroachment.

There is a wide array of trees and shrubs for greenbelt planting. Any programme of this sort, however, should be well designed in advance, planned for a number of years and carefully implemented.

Roadside Plantations

Roadside plantations have several objectives:

- Trees increase the comfort of travellers by providing shade and attractive surroundings;
- Trees may protect the road itself against moving dunes or act as a windbreak for adjacent fields.
- Trees may become an important factor by alleviating timber and fuelwood shortage. In fact, roadside trees are frequently considered a part of the national forest planting programme. Such trees may produce edible fruit, yield pods for feeding animals, furnish food and shelter for birds or, when in bloom, be valuable in beekeeping.

Species should be carefully chosen. Among the important factors which should be considered in planning the use of trees along highways:

- Selection of the species for hardiness, longevity, freedom from windthrow and breakage, attractive appearance and minimal maintenance. In the arid zone there are a large number of native small trees and large shrubs which can be used, as well as some exotic species. Where the environment is suitable, consideration should be given to small patches of deciduous species with colourful foliage.

- Suitability of the species to the climate, topography and soil.
- Location of the trees in relation to road formation. Firstly consideration should be given to the existing road formation so that trees ar not planted close to the inside of curves or near road junctions where they could obscure vision and so create a driving hazard. Secondly, consideration should be given to the possibility of the future widening of roads, including the development of double traffic lanes.

Rail-side Plantations

A trend towards rail-side planting for the provision of greenery, protection from dust and winds and creation of additional tree resources has developed in recent years in many countries. This trend is likely to spread to other countries due to the favourable results already achieved in certain countries. Three to six rows of trees on either side of the track are considered useful. The planting techniques are similar to those for roadside planting. Species vary and depend on the prevailing climatic conditions, mainly temperature, soil and rainfall.

Under very arid conditions, the choice of species is rather limited. Where water is available, several species can be selected. Within an area each row can be given over to one or more species. Mixtures are thus created and are considered better as they yield different produce to meet local needs.

CONIFEROUS FORESTS AND TEMPERATE BROADLEAF

Of closed canopy forests, perhaps none has been as extensively utilised and altered by humans as the temperate forests. Significant portions of Europe and eastern Asia that once supported forests have long since been converted to pasture and agriculture, and little of the original vegetation remains. Similarly, much of the temperate forest in the USA has been cut, although a large percentage returned to forest after cutting or agricultural abandonment. None the less, forests still occupy moderate to significant portions of the temperate regions of North and South America, Europe, Asia and Australia, and are of significant economic and ecological importance.

Distribution and Extent

There are about 1.1 million km^2 of temperate broadleaf deciduous forests in the world, with perhaps another 3-8 million km^2 of temperate coniferous and broadleaf evergreen forests and woodlands. In North America, temperate broadleaf forests are confined primarily to the eastern USA and south-east Canada. West of the central prairie region are the interior and mountainous coniferous forests, and beyond them the coastal temperate coniferous rain forests of the Pacific North-west. Latitudinally, they occur between the subtropical zone (about 28-30°N) and the boreal zone (46-47° N), although the temperate coniferous rain forests of the Pacific coast of Canada and Alaska grow a little north of 56° N. Smaller areas of temperate forest occur in southern Mexico and into Belize and Guatemala, principally at high elevations and along river courses at 0-2500 m.

The South American temperate forests grow primarily along the far south-western coast and the southernmost tip of the continent. The deciduous forests east of the Andes grow south of 45° S, while the coastal rain forests grow between 41 and 56° S. Temperate broadleaf evergreen forests grow principally in Chile, with a minor extension into Argentina. European temperate forests occur predominantly in western and central Europe, continuing in a narrow band through eastern Europe to the Ural Mountains in Russia. Temperate deciduous forests also occur in the Near East in a band around the southern Caspian Sea, broadening to the east into the Iranian highlands, Turkmenistan, Uzbekistan and Kazakhstan. Mountainous areas are forested primarily by conifers.

Asian temperate deciduous forests are found primarily in the central and south-eastern region, including mainland China, Taiwan, Korea and Japan. The region lies between about 26° and 50° N, with conifers increasing in importance with latitude and elevation. Temperate forests, primarily coniferous but often with abundant deciduous canopy species, also occur in central Asia, principally at higher elevations in the Himalayan region. In Australia, moist temperate deciduous forests are found predominantly in northern and western Tasmania, while the mainland is occupied principally by the evergreen *Eucalyptus* forests of the south-east and southwest. Much of New Zealand is temperate broadleaf evergreen forest.

Soils

Soils are naturally variable, although the dominant soil orders found under broadleaf temperate forests are Alfisols, Inceptisols and Ultisols, with substantial areas of Entisols in floodplain environments (Archibold 1995). Spodosols and Ultisols are most common under temperate coniferous forest, but ash-derived Andosols are common in volcanic regions. Alfisols, Inceptisols and some Ultisols are relatively fertile, with high base saturation. Soil pH typically ranges from slightly acidic to slightly basic, although under conifers and some hardwoods, or in low-lying areas with impeded drainage and decomposition, soil pH may be 4 or less.

Climate

As expected for a forest type distributed so widely, prevailing climatic conditions vary substantially. Generally, though, temperate environments are characterised by warm, relatively humid summers and cool-cold winters. Where precipitation is insufficient to support closed canopy forests, temperate forests variously grade into grassland (steppe or prairie), savannah or open woodlands. To the north, where both temperature and precipitation decline, they merge with the boreal forests, and to the south with (sub)tropical vegetation.

At lower elevations, mean temperature during the coldest month typically ranges between -5 and 10°C, with mean temperatures between 10 and 18°C. In mountainous regions, the temperatures can be significantly lower, approaching those of the more northerly boreal zone. Mean annual temperatures typically range between 8 and 13°C but can be lower or higher, depending on geographical location and local elevation.

Precipitation is year-round, although there are significant geographical differences in seasonal distribution. Winters are wetter than summers in much of the temperate forest region, although precipitation is more evenly distributed in portions of eastern North America, western Europe, far southern South America and New Zealand. In the temperate coniferous rain forests of the north-western USA and southeastern Canada, summers are dry and most precipitation falls during the late autumn and winter. Mean annual precipitation is less than 500 mm in some portions of South America and Russia. At the other extreme are the temperate coniferous rain forests of the north-west North American Pacific coast (cover photo), where annual precipitation approaches 2500 mm. More typical values for the temperate forests of eastern North America, western Europe and South-East Asia are 550-1300 mm.

Regional Formations

Although there are distinct regional compositional differences, there is remarkable taxonomic similarity among the Asian, European and North American formations. Numerous genera are common to all three, including *Acer, Betula, Fagus, Fraxinus, Populus, Quercus* and *Ulmus.* The flora of North America and eastern Asia are most similar to one another, and share several genera of forest trees not found in Europe, including *Catalpa, Diospyros, Liriodendron, Nyssa* and *Sassafras.* There are further such affinities in the herbaceous flora of the forest floor (Li 1972).

Although the modern floras evolved under similar climatic regimes, the taxonomic similarity is due primarily to their common geological past. Both the North American and Eurasian continents were connected until the mid-Tertiary, after the period (Cretaceous) during which most of the genera common to the temperate forests of the Northern Hemisphere differentiated. The South American and Australasian formations do not share a common geological past and differ tax-onomically from those of the Northern Hemisphere. However, they do share some families (*e.g.* Fagaceae). Of the three Northern Hemisphere formations, the Asian is the richest, followed by the North American and the European distantly third. The comparatively depauperate European temperate flora is believed to have resulted largely from the east-west orientation of the region's mountain ranges, which inhibited migration and reduced available refugia during past glaciations.

Asian

On the Asian mainland, temperate deciduous forests occupy a coastal band along the China Sea from the Korean Peninsula and into China in the vicinity of Tianjin. From here, deciduous forests broaden throughout the Chinese lowlands to the west (to the Mongshan Mountains) and south to the Yangtze River. Intermixed along the Chinese coast and in the interior highlands are mixed broadleaf-coniferous forests.

Further to the south, evergreen broadleaf trees become more abundant. Coniferous forests become more prominent in the northern parts of both China and Korea. Japan's forests consist primarily of temperate broadleaf deciduous communities in the central-north region, with conifers more abundant at higher elevations and latitudes. To the south, between about 26° 30' and 35° 30'N, lie the temperate broadleaf evergreen forests. Ching (1991) has described the composition of the region's forest in some detail and I only briefly summarise them here.

China. In China's north-eastern temperate forest region (eastern Heilongjiang and Liaoning provinces), conifers occur at the highest elevations and the northern latitudes, lending the forests a (sub)boreal character. Below these forests (up to 1100m elevation) are the *Pinus koraiensis-mixed* hardwood forests, some of which are rather species-rich. Subtypes include the mixed hardwood-*Pinus koraiensis* forest, oak forests *(Quercus mongolica)* and *Betula* forests, the latter of which occur at higher latitudes and elevations.

China's northern temperate forest is roughly triangular in shape, extending from 32° 30' to 42° 30'N and from 103° 30' to 124° 10'E. The major types within the region include the northern oak forest and the southern oak forest. Common throughout the northern oak forest is *Quercus liaotungensis,* with increasing *Quercus mongolica* to the east.

The Liaotung hills and plains comprise the central portion of the northern oak forest area, with forest composition varying by landform and geography. In the mountainous regions of the Hebei and Laioning hills, broadleaf trees grow to elevations of about 1600 m, the forests being dominated by *Quercus* spp. along with *Betula* and *Populus.* The forest of the high loess plateaus of Shaanxi and Shanxi consist of *Pinus* and various hardwoods. The *Picea meyeri* and *Picea wilsonii* characteristic of the original primary forest have been widely replaced on the plateau by *Betula* in disturbed areas.

The southern oak forest of the northern region includes not only deciduous species but also broadleaf evergreens. It is distinguished from the northern oak forest by the scarcity of *Quercus liaotungensis* and the presence of the evergreen *Quercus variabilis.* The vegetation of eastern Shandong bears close similarity to many of Japan's forests and is floristically diverse. Frost-tolerant *Tilia* spp. and *Corylus* spp. are found to the north, and *Machilus thunbergii* is present further south. The central and southern Shandong hills are forested with *Quercus*-dominated communities, as are the plains and hills of southern Shanxi and central Shaanxi. *Betula* and *Salix* become progressively more important with elevation, as do the conifers *Pinus armandii, Pinus tabulaeformis* and *Abies fargesii.*

The Yangtze River drainage in east central China is a transitional zone between the temperate forests to the north and the subtropical forests to the south. There is a high degree of endemism. The genus *Quercus* is well represented, and the tree flora is a mix of more northerly and southern species. Deciduous species and conifers are intermixed with the broadleaf evergreens, which include the genera *Actinodaphne, Castanopsis* and *Lindera.* Bamboo 'forests' grow here, including *Aruninaria* spp. and *Phyllostachus* spp. The upper river valley is particularly diverse, and includes 50 broadleaf and 12 coniferous genera.

Forest of *Abies, Picea* and *Tsuga* occupy the high elevations.

Korean Peninsula. There is disagreement over the classification and latitudinal zonation of Korea's forests but the following types are relatively distinctive: warm temperate broadleaf evergreen, southern deciduous, northern deciduous and cool temperate/boreal coniferous.

The warm temperate broadleaf evergreen forests occupy the southernmost coast of the Korean Peninsula, and include *Quercus* spp., *Castanopsis* spp., *Cinnamomum camphora* and *Machilus thunbergii.* Temperate deciduous forests lie north of approximately 35° N and account for about 85 per cent of Korea's forested land base. Species common to the southern deciduous forests are *Acer formo-sum, Carpinus laxiflora, Carpinus tschonoskii, Quercus mongolica* and *Quercus serrata.* The northern deciduous species include *Celtis koraiensis, Choseia bractesa, Hemiptera dividii, Larix olgensis, Magnolia parviflora* and *Populus maximowiczi.* Other species found variously in the temperate region include *Acer mono, Betula* spp., *Carpinus* spp., *Fraxinus* spp. and numerous *Quercus* spp. In the plateaus and mountains of northern Korea, coniferous species dominate the overstorey, including *Abies nephrolepis, Abies holophylla, Larix dahurica* and *Picea jezoensis. Pinus pumila,* a dwarf species, grows at the highest elevations (above 1200-1600 m). Broadleaf understorey species include *Acer* spp., *Betula* spp., *Tilia* spp. and *Ulmus* spp.

Japan. The forests of Japan include warm temperate broadleaf evergreen, temperate deciduous and cool/cold coniferous. The warm temperate broadleaf evergreen forests occur on the islands of Honshu, Shikoku and Kyushu, and extend from the lowlands to 600-850m. Species characteristic of the region include *Castanopsis cuspidata, Cycobalanus* spp., *Cinnamomum* spp., *Machilus* spp. and *Quercus* spp. In more mature forests, *Quercus gilva* grows on deeper soils, *Quercus slicina* on steep slopes and *Quercus glauca* over limestone. Following disturbance, *Quercus acutissima, Quercus serrata* and *Quercus variabilis* establish in mid-succession. The conifers *Abies firma, Podocarpus macrophyllus, Podocarpus nagi, Torreya nucifera, Pinus thunbergii* and *Tsuga sieboldii* variously occur, depending on latitude and proximity to the coast.

Temperate deciduous forests on the four main Japanese islands (Hokkaido, Honshu, Shikoku and Kyushu) are diverse in species and habitat, too much so to adequately describe here. In general, though, *Fagus crenata* dominates many forests. Its associates vary, depending on topography, disturbance history, soils, etc. Sites facing the Sea of Japan are commonly cold, snowy and windy, and species like *Acer tschonoskii, Acer matsumarae, Alnus maximowiczii, Quercus mongolica* and *Sorbus commixta* are common.

Various species of *Alnus, Populus* and *Salix* grow in floodplain environments, and the conifers *Abies sachalinensis, Abies firma, Abies mariesii, Picea jezoensis, Pinus thunbergii* and *Tsuga sieboldii* variously grow with species of *Betula, Carpinus, Fraxinus, Magnolia, Quercus, Tilia* and *Ulmus* in montane regions. *Quercus cripsula* is also a common dominant, particularly on Hokkaido, and grows well on somewhat drier sites than *Fagus crenata.* Its associates include *Abies sachalinensis, Acer mono, Betula ermanii, Kalopanax pictus, Ostrya japonica, Picea jezoensis, Tilia japonica* and *Ulmus davidiana* var. *japonica. Pinus densiflora* sometimes forms pure stands, with a broadleaf understorey. At higher latitudes and altitudes, the cool/cold coniferous forests are dominated by *Picea* spp. and *Abies* spp. Broadleaf associates include *Betula* spp., *Fraxinus mandshurica, Kalopanax septemlobus* and *Fagus crenata.*

Australasian

Australia and Tasmania. The majority of Australia's forests are considered (sub)tropical, some of which (the south-western and eastern *Eucalyptus* forests) have been described above. Webb and Tracey (1994) consider Australian forests south of about 38° S to be temperate, and I adopt that latitudinal boundary here. However, others consider the temperate forest zone to extend as far north as about 25° S, and in reality the transition from (sub)tropical to temperate is characteristically indistinct.

Regardless of the precise line of demarcation, the temperate forests of the Australian mainland are dominated primarily by *Eucalyptus,* and account for approximately 29 million ha of a total 42 million ha of closed forest, both temperate and tropical. Groves (1994) has described the dominant vegetation in some wet sclerophyll Australian temperate *Eucalyptus* forests distributed among four main geographical regions. Similar tabulations, including both wet and dry sclerophyllous *Eucalyptus* forests divided into major alliances, have been made by Ovington and Pryor (1983). Despite the diversity of *Eucalyptus,* temperate forests are often dominated by only a few species, for example *E.*

obliqua-E. viminalis forests in parts of Victoria and *E. marginata-E. calophylla* forests in Western Australia.

Temperate rain forest occupies small areas of south-eastern Australia, principally in the territories of Victoria, the north-eastern corner of New South Wales and far south-eastern Queensland. The canopies of Victorian forests are dominated by *Nothofagus cunninghamii,* while *Athersperma moschatum, Acacia melanoxylon, Eucriphia lucida* and *Phyllocladus asplenfolius* are occasional or common associates. *Nothofagus cunninghamii* does not grow in easternmost Victoria, and *Athersperma moschatum, Eleaocarpus holopetalus* and *Telopea oreades* are most common. Eucalypts dominate the drier uplands in the region. In the forests of New South Wales, *Eucryphia moorei, Nothofagus moorei* and *Doryphora sassafras* are common, with the former two species occurring at somewhat higher elevations.

Western Tasmanian temperate rain forests are dominated by *Nothofagus cunninghamii,* which may contain any of about 70 associates, only a few of which typically occur together in a stand. *Athrotaxis* spp. (especially *A. cupressoides)* occur at altitudes greater than 900 m. In drier environments, like eastern Tasmania, *Eucalyptus* spp. are dominant; common ones include *E. delegatensis, E. dalrym-pleana, E. gunnii, E. archeri, E. coccifera, E. urnig-era* and *E.pauciflora.*

New Zealand. The climate of New Zealand ranges from subtropical to cool/cold temperate, but here I consider it mainly temperate. Generally, the native taxa are similar to those of the South American and Tasmanian forests. Most are evergreen in habit, including both conifers and broadleaf species. There are two broad categories of New Zealand forest: conifer-broadleaf, and beech.

The conifer-broadleaf forests are predominantly lowland and are dominated by podocarps, including the genera *Dacrycarpus, Dacrydium, Phyllocladus* and, of course, *Podocarpus.* Also present are two species of *Libocedrus* and the kauri *(Agathis australis).* The warmer inland forests support primarily *Podocarpus spicatus* and *Podocarpus totara* as canopy dominants, along with a more minor component of *Podocarpus dacrydiodes* and *Dacridium cupressinum.* Hardwood associates of lower stature include the genera *Beilschmiedia, Knightia, Laurelia, Litsea* and *Nestegis.* Forests range from relatively open to closed canopy, depending on species and habitat. For example, *Dacridium cupressinum, Podocarpus hallii* and *Podocarpus totara* are most common on drier sites and may form near-monospecific stands, while *Podocarpus dacrydiodes* often dominates on wetter, swampy sites. In the coastal hills and plains, *Dacridium cupressinum* is commonly dominant, although *Podocarpus hallii* and *Podocarpus totara* are likely to occupy more xeric sites.

Genera common and often dominant in coastal lowland forests include *Elaeocarpus, Met-rosideros* and *Weinmannia*. The conifer-broadleaf shrub community in these forests is often well developed, with abundant species. The overstorey tree *Weinmannia racemosa* is abundant and widely distributed on the wetter, western side of both islands. Another important tree species is the giant kauri *(Agathis australis),* much reduced in extent and growing primarily north of 38° S, particularly on ridges. It grows with podocarps and various hardwoods, and occurs both singly and in groups or thickets on poorer soils.

The beech, or *Nothofagus,* forests occupy cold, wet, montane environments and are characteristically lower in stature and less species-rich than the conifer-broadleaf forests. Four species of *Nothofagus* grow in New Zealand: *N. solandri, N. menziesii, N. fusca* and *N. trun-cata. Nothfagus solandri* is the most widespread of the four species and there are two recognised varieties: var. *solandri* and var. *cliffortoides.*

The former occupies warm, dry lowland environments primarily from East Cape on North Island to the centre of South Island, while var. *cliffortoides* grows at higher altitudes or on poor soils under high rainfall to the south. *Nothfagus solandri* forms nearly pure stands, with a sparse under-storey composed of shrubs (especially *Coprosma* spp.), lichens and mosses, although both varieties co-occur where their geographical ranges overlap. *Nothofagus fusca* is distributed from East Cape to the southern end of South Island, and grows best on deeper, well-drained soils. *Nothofagus menziesii* has similar site requirements and dominates the subalpine zone of western South Island and the eastern mountain ranges of North Island. *Nothofagus truncata* grows on drier, less fertile sites, primarily north of 42° S.

Near Eastern

The Near Eastern temperate forest lies between 35 and 45° N, from about 27° E in eastern Turkey to about 58° E in eastern Iran. The forests of this region are a mixture of both European and Colchian species, with elements of (sub)Mediterranean and steppe flora.

In western Turkey, oak forests are common, composed of various species of *Quercus* and other hardwoods. *Fagus orientalis* grows at higher elevations, though not in abundance. In the western portion of the Anatolian plateau in Turkey, *Quercus*-dominated temperate broadleaf forests grow between 1000 and 2000 m and contain a number of other genera as well. Extending into Iran further to the east are other forests dominated by *Quercus.* Various species of *Acer, Pyrus* and *Prunus* also occur, as well as a few species, like *Fraxinus excelsior,* more commonly found in the Middle European and Submediterranean regions. On alluvial soils at lower elevations (<400m) grow more diverse forests, which in addition to *Quercus* spp. include *Alnus barbata* and *Carpinus* spp. Moist forests between the Caspian Sea and Black Sea contain a number of species, including many missing from the European forests to the west. Further east, the species and assemblages change but a few genera predominate, with a number of low-stature woody species occupying the understorey (*e.g. Buxus, Crataegus, Corylus, Ilex, Ligustrum, Rhododendron* and *Sambucus).*

At high elevations (600-1100 m) are the beech *(Fagus orientalis)* forests. Near pure *Fagus orientalis* forests grow between the Sakarya River in western Turkey and the Turkey-Georgia border. Between 600 and 1300 m on calcareous northern slopes on the mountains of the southern Crimean Peninsula, *Fagus orientalis* grows alone or with *Carpinus betulus, Fraxinus excelsior, Tilia cordata* and *Ulmus glabra.* On southern slopes, it is found with *Acer hycranum, Pinus nigra, Carpinus orientalis* and *Quercus petraea.* In the Caucasus Mountains, it grows between 1000 and 1500m with *Acer* spp., *Fraxinus excelsior, Tilia platyphyllos* and *Ulmus glabra.* At higher elevations, *Fagus orientalis* grows with coniferous species like *Abies born-mülleriana, Abies equitrojani, Abies nordmanni-ana, Picea orientalis* and *Pinus sylvestris,* along with some of the broadleaf species found at the lower elevations.

North and Central American

Temperate forests occupy much of North America, from north of the border between the USA and Canada south into Mexico and some Central American countries at high altitudes. The conifer-dominated forests of western North America are separated from the more deciduous forests of the east by vast areas of former prairie now dominated by agriculture where rainfall or irrigation permit.

USA and Canada. The eastern North American temperate forest is primarily deciduous, although the forests of the south-eastern USA contain large amounts of conifers, both natural and planted, as do the states bordering the Great Lakes. The forests of the region have been extensively described, especially by Braun (1950) and more recently Barnes (1991). I follow Barnes's (1991) classification scheme, which is derived from Braun's (1950) and Küchler's (1964). The South-eastern Evergreen region covers the coastal plain of the south-eastern USA, stretching roughly from Virginia along the Atlantic coast, continuing south across Florida and the Gulf of Mexico to Texas. *Pinus palustris-Aristida stricta* savannahs formerly dominated much of the uplands but have been largely converted to other uses or cover types, and both *Pinus elliottii* and *Pinus taeda* are presently the most abundant conifers in the region. *Magnolia grandiflora* and *Fagus grandifolia* are believed to be climax dominants on mesic sites, although a number of other genera are abundantly represented.

Inland of the South-eastern Evergreen region is the Oak-Hickory-Pine region, extending from Pennsylvania to Alabama and bordered by the coastal plain towards the Atlantic Ocean and the Appalachian mountains to the west. As its name implies, the principal species include those in the genera *Quercus* and *Carya,* often in mixture with species common to adjacent regions, including *Pinus.*

The Appalachian Oak region includes the Appalachian mountains, from Georgia north and eastward into eastern New York and Massachusetts, and is the most floristically diverse in North America. Oaks are abundant at all elevations and aspects. At higher elevations (> 1100 m), species are more characteristic of the beech-sugar maple and sugar maple-basswood forests to the north. At the highest elevations are boreal forests composed of *Picea rubens* and the endemic *Abies fraserii* in the south, and *Picea rubens* and *Abies balsamea* to east to south-west from western Pennsylvania to northern Alabama, and coincides physiographi-cally with much of the unglaciated Appalachian plateaus. The vegetation comprises principally species from the regions which it borders. The Western Mixed Mesophytic region lies immediately to its west, covering central and western Tennessee and Kentucky, the southern boundary of the Wisconsinan Glaciation in southern Indiana and Ohio, and northern Mississippi and Alabama. Forests dominated by *Quercus* spp. cover most of the region, with a large component of *Carya* and *Pinus.*

The westernmost region is the Oak-Hickory, bordered to the west formerly by prairies and presently by agricultural lands. It extends from Manitoba, Canada to southern Texas. South of the glaciated portion lie the Ouachita Mountains and Ozark Plateau. *Quercus alba* is the dominant species on dry-mesic uplands, with various species of *Carya, Quercus* and *Pinus* characteristic of drier sites. In the glaciated region to the north, *Quercus* spp. are common on drier uplands, while *Fagus grandifolia* and *Acer sac-charum* grow on more mesic sites. *Quercus macrocarpa* formerly formed savannahs interspersed with prairie, most of which has been lost to agriculture and urbanisation.

The Sugar Maple-Beech region occurs on glaciated till in the southern half of Michigan's lower peninsula, Indiana and Ohio; *Acer saccha-rum* and *Fagus grandifolia* are the dominant species in mature forests, although a number of other species are common. The Sugar Maple-Basswood region to the north of the Sugar Maple-Beech region is situated on glaciated terrain and, as the name implies, *Acer saccharum* and *Tilia americana* are dominant. Other associates are those characteristic of the Sugar Maple-Beech region.

The Northern Hardwood-Conifer region extends across the northern USA and southern Canada from Manitoba to Maine and Nova Scotia. The terrain is glaciated and supports a wide array of forest types. Species common to the Sugar Maple-Beech region predominate on glacial till, although the region extends beyond the northern range of *Carya, Liriodendron, Nyssa* and other genera common to the more southern forest regions. *Quercus ellipsoidalis* grows on dry glacial outwash plains, but fire-prone *Pinus banksiana* is the most dominant species on these soils. Much of the presettlement forest was dominated by *Pinus strobus* but virtually all of the original forest was cut and converted naturally to hardwoods, including *Betula papyrifera, Populus tremuloides, Populus grandidentata* and *Quercus* spp. Coniferous swamps on calcareous soils often consist predominantly of *Thuja occidentalis* with lesser amounts of other hardwoods.

The temperate forests of western North America differ from those in the east by a preponderance of conifers rather than hardwoods. Furthermore, much of forested western North America is mountainous and the vegetation is often distributed in rather discrete elevational bands, as opposed to the more subtle changes with landscape position associated with the eastern deciduous forests.

The Rocky Mountains extend from north-central Canada, through the USA and into Mexico, and form the eastern border of North America's temperate coniferous forests. At the lowest elevations (1500-1600m), *Pinus edulis* and *Pinus monophylla* grow with various species of *Juniperus* and *Quercus*. At higher elevations are *Pinus ponderosa* forests, both closed canopy and open park-like stands, above which lie forests dominated by *Pseudotsuga menziesii*. In the southern Rockies, the associates of *Pseudotsuga menziesii* are few, but in the northern Rockies it is variously found with *Abies grandis, Larix occidentalis, Pinus contorta* and *Populus tremuloides*. With increasing elevation in the northern Rockies, *Pseudotsuga menziesii* occurs with *Thuja plicata* and *Tsuga heterophylla*. These two species often form nearly pure stands. *Populus tremuloides* grows at similar elevations, colonising after disturbances, principally fire. Just below the treeline are found *Picea engelmannii* and *Abies lasiocarpa*. In the far northern Canadian Rockies, the principal sub-alpine species are *Abies lasiocarpa, Picea glauca* and *Pinus contorta*.

In the Sierra Nevada Mountains of California, *Pinus sabiana* is a dominant conifer in the western foothills, growing with *Quercus* spp. On western slopes, *Pinus lambertiana* grows with *Pinus ponderosa* at higher elevations (750-1800 m), together with *Abies concolor* and *Calo-cedrus decurrens*.

Pinus jeffreyi occupies lower eastern slopes and higher west-facing elevations. Further south, *Pseudotsuga macrocarpa* and *Pinus coulteri* grow in the elevations between desert scrub and the mixed conifer forests described above. *Sequoiadendron giganteum*, the world's largest trees, are distributed between 1350 and 2250 m on the western slopes of the Sierra Nevada. *Abies magnifica* and *Pinus contorta* grow between roughly 1800 and 2400 m, in regions of extremely high snowfall (4-20 m snow-pack). The subalpine forests (>2400m) consist of *Pinus albicaulis, Tsuga mertensiana, Pinus balfouriana, Pinus flexilis* and the long-lived (>4000 years) *Pinus aristata*.

The coastal mountain ranges of the northwestern USA and south-west Canada differ from the interior mountains principally by the abundant precipitation they receive (upwards of 2000 mm annually) and the exceptionally large size that many of the species attain. Along the Pacific coast from northern California into southeastern Alaska and the Kodiak Islands, *Picea sitchensis* forests are found in narrow bands (generally only a few kilometres wide) below 150 to 600 m elevation. Its associates include *Pseudotsuga menziesii, Thuja plicata, Tsuga heterophylla* and *Alnus rubra*. In southern Oregon and northern California, *Sequoia sempervirens* forests grow at similar elevations, extending as far as 15-20 km inland, and growing on slopes and sometimes at higher elevations.

Above the *Picea sitchensis* zone is the *Tsuga heterophylla* zone (550-1200m), above which *Abies amabilis* dominates the elevational bands ranging between 600 and 1500 m. The highest elevation forests are the subalpine *Tsuga mertensiana* forests that grow as high as 2000 m in southern Oregon. Principal associates include *Abies amabilis, Abies lasio-carpa* and *Abies procera. Tsuga mertensiana* is replaced by *Abies lasiocarpa* in the interior portions of the Cascades and on the drier eastern slopes. The most extensive forests of the eastern slopes of the Cascades are those dominated by *Abies grandis,* which grow at elevations of 1000-1500m in the north and 1000-2000m in southern Oregon.

Central America and Mexico. Geographically speaking, the forests of the region are primarily (sub)tropical in nature, although there is a large degree of taxonomic similarity between the vegetation of the USA and Mexico and Central America. Some of the similarity is due to the continuity of species' range distributions, but it is also believed that the region provided a refuge during cool periods of the Oligocene, after which geological uplift and further climatic changes isolated the temperate forests to disjunct islands among otherwise (sub)tropical forests from Mexico southward into Nicaragua.

Among deciduous species, *Carpinus carolini-ana, Liquidambar styracifl.ua, Nyssa sylvatica, Ostrya virginiana* and *Prunus serotina* grow in both the USA and Mexico. Species with a high degree of similarity include *Fagus mexicana* and *Fagus grandifolia, Carya ovata* var. *mexicana* and *Carya ovata,* and *Acer sutchii* and *Acer saccha-rum.* Several genera are well represented in both regions, including *Alnus, Carya, Fraxinus, Juglans, Magnolia, Platanus, Populus* and *Salix.* The genus *Quercus* is especially abundant, both deciduous and evergreen.

In montane deciduous forests, *Juglans olancha, Liquidambar styraciflua* and *Ostrya virginiana* are dominant, growing at 1200-1800m in the Sierra de Chucharas in the Mexican state of Tamaulipas. Forests dominated by *Quercus* spp. and *Pinus* spp. often occupy the highest elevations (> 1800 m), with (sub)tropical species prominent in the lower elevations. Riverine forests may be composed primarily of deciduous genera like *Carpinus, Fraxinus, Platanus, Populus, Salix* and *Ulmus,* but uplands typically contain representatives from more (sub)tropical families like Clethraceae, Fabaceae, Sabiaceae, Lauraceae, Staphylaceae, Cunoniaceae and Rutaceae.

Mexico and Central America also contain a rich coniferous flora, especially of pines. There are at least 47 *Pinus* species, which is about half the world total. Many species are five-needled pines, and they occupy niches from coastal to high montane environments. Some, *e.g. Pinus caribaea* and *P. patula,* are widely planted in the tropics.

European

Temperate forests once covered much of Europe and the British Isles, but most of the original forest has long since been converted to agriculture. Moreover, many introduced species have been extensively planted, for example plantations of the western North American conifers *Picea sitchensis* and *Pseudotsuga menziesii* in the British Isles. None the less, temperate forests remain an important resource and there are numerous recognised community types. Jahn (1991) has extensively described the European deciduous forests, and I make considerable use of his excellent work in my more abbreviated summary here. Two regions of temperate forest can be recognised: the Middle European, which occupies a band between about 43° and 55-60°N, and the Submediterranean, which generally follows the southern border of the Middle European south to a latitude of about 40° N.

Middle European. In the Middle European region are four somewhat distinctive provinces: the Atlantic, Subatlantic, Central European and Sarmatic. Throughout the region are genera common to most of the temperate forest of the Northern Hemisphere, including *Acer, Betula, Carpinus, Fraxinus, Pinus, Populus, Prunus, Quercus, Sorbus, Tilia* and *Ulmus*. The genus *Fagus* is represented principally by *Fagus sylvatica,* which is distributed throughout most of the Middle European region and is, alongside *Quercus,* the most conspicuous and perhaps most ecologically important tree in Europe.

The Atlantic province includes the British Isles and the coastal regions of western Europe, and a number of characteristic native species are, or were, widely distributed. In the British Isles, *Quercus* spp. still dominate many sites, although *Acer* spp., *Carpinus betulus, Fraxinus excelsior, Tilia* spp., and other species are abundant. *Fraxinus excelsior* is often the dominant species on high pH soils, with *Quercus robur* and *Quercus petraea* on sites of lower pH and fertility. *Fagus sylvatica* and *Fagus sylvatica-Quercus petraea* woodlands occur in south-eastern England. *Fagus sylvatica* is also common in north-western France, portions of Belgium and throughout the Middle European and portions of the Submediterranean regions. In the Atlantic province, *Fagus sylvatica* is typically absent only on sites that are wet, steep, calcareous or very sandy. The Subatlantic province is orientated along a north-east to south-west axis, covering an east-west region between about 8 and 15°W in southern Fennoscandinavia and about 3° W and 3°E in southern France and northern Spain. *Quercus robur* is common in the north and co-occurs with a number of associates depending on soil conditions. *Fagus sylvatica* is prominent on upland sites in the north and in the hills and mountains to the south, and forms a complex regional mosaic of communities, determined by a combination of climatic, soil and physiographic factors. In the southern portion of the Subatlantic province, *Quercus* spp. become increasingly prominent, particularly at lower elevations. In the eastern Pyrenees, *Quercus pubescens* is found at the lowest elevations, *Fagus sylvatica* and *Abies alba* at higher elevations, with *Abies alba* increasing in importance with altitude.

The Central European province covers the former East Germany, western Poland and western Czech Republic and is characterised by a more continental climate than the Atlantic or Subatlantic provinces. *Fagus sylvatica* reaches its easternmost extent here, and occurs extensively north of the Baltic Sea. It has many associates, depending on elevation, and often has a more species-rich ground layer than comparable forests in the Subatlantic province. *Carpinus betulus-Quercus petraea* communities formerly occupied rich, mesic plains and uplands, but most have been converted to agriculture. *Betula-Quercus* forests grow on nutrient-poor, low pH soils, with *Betula pubescens* and *Quercus robur* common on wet sites and *Betula pendula* and *Quercus petraea* on drier sites. *Pinus sylvestris* and *Picea abies* are locally intermixed in the north and east, with *Larix decidua* common to the hills of southern Poland.

The Sarmatic province extends in a band eastward to nearly 60° E, its north-south boundaries tapering from about 50-60° N in the west to about 53-54° N at its eastern edge. To the south lies steppe and to the north boreal forests. Generally, the Sarmatic province is characterised by a progressive west-to-east loss of the deciduous species characteristic of the three provinces to the west and an increase in the importance of conifers. For example, there is a tendency for nutrient-poor sites to shift from domination by *Betula* and *Quercus robur* in the Subatlantic province to domination by *Pinus* in the Sarmatic. Similarly, on more fertile sites, *Fagus sylvatica* is no longer present, and has been replaced by *Tilia, Quercus robur, Carpinus* and *Pinus. Pinus sylvestris* is distributed throughout the Sarmatic province and is an associate or dominant on all but the most fertile, moist soils. On sandy soils it forms nearly pure stands, without deciduous associates. *Picea abies* is also common, occurring with *Pinus sylvestris,* and increasing in importance with soil fertility and moisture. *Abies alba* occurs in the eastern part of the province, with *Abies sibirica* in the far western portion. *Populus tremula* and *Quercus robur* are present along the southern boundary, although the latter is uncommon in the north. In contrast to the northern portions of the Sarmatic province, the south is largely deciduous. *Quercus robur* and *Tilia cordata* are dominant, but *Carpinus betulus* is an important (though often low stature) associate.

Intermixed or adjacent to these four provinces of the Middle European region are the forests of the Alps and the Carpathian Mountains. The Western Alps (orientated north-south and running from the Mediterranean to Grenoble, France) are occupied by *Quercus pubescens* forests in the west-facing foothills, with *Fagus sylvatica, Quercus petraea* and *Pinus sylvestris* occurring at higher elevations. High-elevation interior forests contain a mix of deciduous and coniferous species. East-facing forests are predominantly deciduous hardwoods, with a large *Quercus* component.

In the Middle Alps, the north-facing foothills are forested with *Quercus petraea* and *Carpinus betulus,* which are replaced by *Fagus sylvatica* and *Acer pseudoplatanus* at higher elevations, *Abies alba* and *Picea abies* in the interior, *Fagus sylvatica* on upper south-facing slopes and *Cas-tanea sativa* and *Quercus petraea* on the south-facing foothills. The Eastern Alps (Verona, Italy to the Dinara mountains of the former Yugoslavia) are occupied by a variety of hardwoods, with an increasing abundance of conifers at higher elevations. Similar species occur on the south-facing slopes, with the addition of *Fraxinus ornus, Ostrya carpinifolia* and *Quercus* spp. in the low foothills and foothill-plains transition.

The Carpathian Mountains extend in an arc from Slovakia east through south-west Ukraine and into central Romania. The forests of the foothills and submontane zone (up to about 600 m) of the western Carpathians are predominantly mixed hardwood forests, with a minor component of *Fagus sylvatica.* On the eastern side, forest and steppe alternate at the lowest elevations, and are progressively replaced by mixed hardwoods, and then coniferous forests, as elevation increases.

Submediterranean. The Submediterranean region is characterised by a preponderance of deciduous oaks *(Quercus* spp.) and represents the southern distribution of many of the species common to the mixed forests of the Middle European region to the north. To the south, where summer droughts are more extended, grow the evergreen Mediterranean forests. The vegetation is quite variable and complex throughout and is sensitive to climatic and site factors in this region, which borders temperate and Mediterranean climatic zones. Deciduous species of the Submediterranean region include *Carpinus orien-talis, Fagus sylvatica, Fagus moesica, Fagus orientalis, Fraxinus ornus, Ostrya carpinifolia, Quercus cerris, Quercus faginea, Quercus frainetto, Quercus pubescens* and *Quercus pyren-aica.* Various species of *Acer* and *Sorbus* are also present. *Fagus* spp. are characteristic of montane regions and co-occur with *Abies* spp. and *Juni-perus* spp.. *Pinus* spp. are present, with *Pinus nigra* growing with *Quercus cerris, Quercus faginea, Quercus pubescens* and *Quercus pyrenaica* in submontane environments, *Pinus sylvestris* common in montane zones, and *Pinus peuce, Pinus heldreichii* and *Pinus mugo* growing at high elevations.

South American

The South American temperate forests are the smallest in extent, confined primarily to Chile within and between the coastal and Andean mountain ranges, with a small extension into Argentina. Both broadleaf evergreen and deciduous species are present, often in mixture, although evergreen species are dominant. *Nothofagus* spp. are probably the most abundant trees; a total of 10 species grow in the region, of which seven are deciduous *(N. obliqua, N. alpina, N. pumilio, N. antarctica, N. glauca, N. leoni* and *N. alessandri)* and three evergreen

(N. betu-loides, N. dombeyi and*N. nitida).* The only other deciduous species present is *Acacia cavens,* which grows in a more Mediterranean climate in the northern part of the temperate region, between about 31 and 38° S.

Nothofagus glauca and *N. obliqua* grow in the wetter climate to the south of the *Acacia* forests in central Chile, the former on eastern and northern slopes and the latter on lowland sites. *Nothofagus alessandri* mixes with *N. glauca* on moister sites of the coastal mountains. *Nothofagus obliqua* dominates much of the lowlands and valleys south of 36-38° S, depending on proximity from the ocean. It is often found with *N. alpina* and *N. dombeyi,* the latter of which often forms pure stands in cool wet gorges.

The conifer *Libocedrus chilensis* is often present on dry slopes opposite *N. obliqua,* and *Laurelia sempervirens* and *Persea lingue* are common, though lower-stature, associates of *N. obliqua.* At higher elevations in the Andes, *N. dombeyi* replaces *N. obliqua. Nothofagus alpina* forests occur in narrow elevational bands in the northern Chilean Andes and the species is widespread in mixed forests of the coastal range, often found with *N. dombeyi* and the associates of *N. obliqua. Auracaria araucana* occurs over small areas of the coastal range and is more widespread in the Andes (37-41° S) above the *N. alpina* forests (600-900m). *Nothofagus pumilo* occupies cold snowy sites, at elevations of 1500-1800 m at about 37° S down to sea level at Cape Horn. *Nothofagus antarctica* is a high-elevation krummholz-forming species that also colonises primary successional habitats (*e.g.* volcanic deposits) and other harsh or exposed sites. *Nothofagus nitida, Podocarpus nubigenus* and *Weinmannia trichosperma* occupy humid wetter sites between 600 and 900 m elevation south of about 40° S. *Fitzroya cupres-soides* is a large-sized tree occurring on wetter sites between about 40° and 43° 30'S in both the Andean and coastal ranges, and at elevations from about 700m to the timberline.

NATURAL FORESTS

The forest area estimates described above include undisturbed forests, forests modified by humans through use and management (or 'seminatural' forests) and forests created artificially by humankind (*i.e.* forest plantations) by afforestation or reforestation. (Afforestation is defined as the establishment of a tree crop on an area from which it has always, or for a very long time, been absent. Reforestation is defined as the establishment of a tree crop on forest land.)

In most industrialized countries, particularly in continental Europe, forests are being managed in such a way that at management-unit level a continuum exists from low-intensity management, involving natural regeneration, through more intensive methods involving some artificial planting to highly intensive methods with complete planting and cultivation; this makes it difficult to isolate figures for natural forest and plantations. The distinction between natural or seminatural forests and forest plantations can more easily be made for developing countries and some industrialized countries such as New Zealand in which forest plantations have been established using introduced species.

Interest in natural forests, particularly their role in the conservation of biological diversity, has led to efforts to compare forests today with what is thought to be their original character and to give complete protection to areas of forests which have had no, or minimal, human interference. Although there are difficulties in identifying the extent of natural forest, compounded by problems of definition, some information exists that can be used as an indication of broad patterns of natural forests in various regions.

An attempt has been made by the World Wide Fund for Nature (WWF) to quantify the area of forests in western Europe that has been relatively undisturbed by humans or which has retained much of its natural character. Their report distinguishes between 'virgin forest', defined as 'forest ecosystems whose characteristics are determined exclusively by natural location and environmental factors... without human influences present or visible any more', and 'natural and ancient seminatural forests', which 'have not been planted or sown by man for the past two centuries' and 'which continue to have a large number of the natural elements'.

The study found that only a small proportion (probably < 1 per cent) of the total forest land in northern and western Europe could be considered as virgin forest, which has arisen since the last glaciation. Almost all was located in Sweden, Finland and Norway, with small areas in Greece, Austria and Switzerland and (according to another author) in France. In eastern Europe, Slovakia and Belarus have considerable areas of virgin forest while Poland and Croatia have small areas. In addition, in northern and western Europe, the WWF report identifies natural and ancient seminatural forests representing 2.1 per cent of the total forest cover (1990) of the 16 countries concerned. There are a further 3 million ha of land in the region in national parks and other protected areas and another 50 000 ha in small forest reserves (mostly for nature conservation and scientific research), whose use is tightly restricted.

The situation in temperate and boreal North America is quite different from that of densely populated Europe and Japan, where use and management of forests for many centuries have left very little of the original forest area untouched. 'Old growth forests', as they are called in North America, still cover extensive areas. On the lands managed by the US National Park Service alone, old-growth forests covered 1.97 million ha in 1988. Although not a direct measure of the extent of old growth forest, it may be noted that the area of forest included in national parks and other protected areas in North America (USA and Canada) was reported to be nearly 49 million ha in 1990.

The FAO Forest Resource Assessment 1990 did not distinguish between undisturbed and disturbed 'natural' forests in developing countries. However, the Forest Resource Assessment 1980 made estimates of the areas of undisturbed closed forests (primary forests and old secondary forests where there had been no logging for the last 60-80 years) and of closed forests that were included in national parks and other protected areas (thus relatively undisturbed, at least in theory).

At that time, these two categories together represented 60 per cent of the total closed forest area in the tropics, a proportion varying from 39 per cent in tropical Asia to 59 per cent in tropical Africa and 69 per cent in tropical America. These different proportions by region reflected a slower development of large-scale harvesting in tropical America compared with tropical Africa and Asia, and also the fact that, in tropical America, spontaneous colonization did not follow in the wake of logging as systematically as in the two other regions due to lower population pressure.

Although the two categories do not match the concept of 'virgin forests' of Europe and of 'old growth forests' of North America discussed above, the sum of the two categories nevertheless gives an indication of the amount of forest disturbance (or management, depending on the point of view of the observer) that existed around 1980 in the humid tropics. Although corresponding estimates for 1990 and 1995 are not available, it is likely that the share of undisturbed forests remains higher in the three tropical regions than in Europe and probably in North America too.

Forest Plantations

Recent data in this chapter are also available in *State of the World's Forests 1999,* which also discusses some of the issues concerned with forest plantations in more depth.

Early Development

By the seventeenth century the decline in the area of native forest in European countries led to the planting of trees, largely to provide alternative sources of timber supply. Sometimes plantations were established for the provision of services or other products than timber. They might be planted as shelterbelts, for dune stabilization, for amenity or for the supply of firewood.

Plantations of native species were established at first in areas where forests occurred naturally, for example France, Germany, England and Scotland, but later also on previously unforested land. In France, for example, planting of *Pinus pinaster* was started on the sand dunes of the Landes and with *Pinus sylvestris* and *Picea abies* on former agricultural land in the Vosges; in Germany likewise agricultural land was planted with Norway spruce *(Picea abies)* in Saxony. In tropical countries such as Myanmar (then Burma), teak *(Tectona grandis)* was planted as a native species in the *taungya* system.

Teak was one of the first exotic forest plantation species to be used, being planted in Sri Lanka and the island of Java (Indonesia) from early in the nineteenth century. Exotic species, such as Douglas fir *(Pseudotsuga menziesii),* Sitka spruce *(Picea sitchensis),* Japanese larch *(Larix kaempferi),* lodgepole pine *(Pinus contorta)* and poplar *(Populus deltoides),* were introduced to Europe during the nineteenth century, when they began to play an increasingly important role in forest plantation programmes. Other exotic species, particularly the eucalypts *(Eucalyptus* spp.) and wattle *(Acacia* spp.) from Australia, were introduced as exotics in tropical and subtropical countries from the middle of the nineteenth century, while *Pinus radiata* was introduced from California to New Zealand and other countries such as Chile from the early years of the twentieth century. Box 1.3 describes in detail the experience with forest plantations in Denmark.

The main reason for the establishment of plantations remained the decline in natural forest area and a scarcity of wood, the conversion of woods and forests to agricultural and grazing use. In recent years increased areas of natural forest have been managed for nature conservation, recreation, wildlife parks, etc., and commercial wood production has been either reduced or eliminated. For example, over 25 000 ha of high-yielding hybrid poplar plantations have been established in the north-western USA between 1992 and 1997 in response to both increased demand for poplar wood for orientated-strand board and decreased supply from public forests. On the other hand, some new areas have become available in European countries as land is taken out of agriculture due to trade and market considerations.

Objectives

Forest plantations are tree crops that are in some, but not all, ways analogous to agricultural crops. They often have a simple structure, at least in youth, and are usually composed of one or a few species (but not varieties as in agriculture, except in a few cases such as the intensively bred poplars) chosen for their fast growth, yield of specified products and ease of management. When established for wood production they have a higher productivity of usable wood than natural forests, but due to the way they are managed they do not, indeed cannot, provide the full range of goods and services that natural, seminatural or even secondary forest can provide.

Although the objective of many plantations is the production of industrial roundwood and/or fuelwood, many are established for environmental protection or other services and some also provide non-wood forest products, such as fodder, various foodstuffs, medicines, etc.

Area

The area of forest plantations throughout the world started to increase in the 1970s as many governments became concerned about wood supplies for industry, and in developing countries fuelwood supplies, and has continued to increase since. However, there are no reliable global figures for plantation areas because forests of native species in several developed countries in the temperate and boreal regions, especially in continental Europe, are frequently regenerated naturally and it is not possible to distinguish those areas where supplementary artificial planting has been done.

Furthermore many countries consider their plantations as 'seminatural' forests over a certain age, because as stands mature the clear initial row layout of the trees is lost and other species naturally regenerate under the canopy. For example, Austria, Czech Republic and Finland, in responding to a questionnaire to collect data for the ECE/FAO Temperate and Boreal Forest Resource Assessment (TBFRA), stated that they had no plantations in their countries as defined by the TBFRA process.

In 1995, an approximate estimate of the area of plantations in developed countries was 60 million ha, comprising 13.7 million ha in North America, 22.2 million ha in the Commonwealth of Independent States, 12.1 million ha in Europe and 13.2 million ha in Oceania (Australia, New Zealand, Japan). The most significant areas of plantations were in the Russian Federation (17.3 million ha, 2.1 per cent of the country's total forest area), the USA (13.7 million ha, 6.3 per cent), Japan (10.7 million ha, 44.4 per cent), Ukraine (4.4 million ha, 46.8 per cent) (FAO/ECE 1996), Spain (1.9 million ha, 14.9 per cent), New Zealand (1.54 million ha, 19 per cent of the total forest area in 1996, 91 per cent of which was *Pinus radiata)* and Australia (1.04 million ha in 1994, 85 per cent of which was softwood species, mainly *P. radiata).*

The estimated 'net' plantation area of 55 million ha in developing countries in 1995 was about 2.8 per cent of the total area of forests in developing countries. In 1980 the net plantation area was assessed at about 40 million ha. It has thus increased by about 15 million ha in 15 years, but even this figure may be liable to error. Many developing countries, especially those with large forest plantation programmes, provided updated information on their present plantation plans to FAO in 1996 and 1997 from which the reported annual rate of new plantations of 3 million ha was derived. Note, however, that this is the reported or planned rate, which may not necessarily have been achieved. Most of the countries with large plantation estates indicated that they intended to double their plantation areas between 1995 and 2010.

It was estimated from reported figures that 57 per cent of the forest plantation area consisted of hardwood species and 63 per cent was established for industrial purposes. Nearly three-quarters of these plantations were in the Asia-Pacific region, where China (21 million ha) and India (20 million ha) dominate, while about 15 per cent were in Latin America and 10 per cent in Africa.

One of the trends in the tropics has been that the proportion of industrial plantations established in large blocks fell from 40 per cent of the total plantation area in 1980 to 35 per cent in 1990. The proportion of smaller plantations established through farm forestry or agroforestry programmes grew in importance during the period 1980-95, particularly in the Asia-Oceania region.

Unfortunately, the figures on small-scale private or community-owned forest plantations are even less reliable than for large-scale plantations. Some of these farm forestry or agroforestry plantations supply industrial wood markets for pulpwood. In many countries, particularly those with limited forest area, planted trees grown outside the formal forest area often provide the bulk of fuelwood, poles, construction wood, utility wood, as well as fodder and other non-wood forest products for household use.

Of the area of hardwood plantations planted for industrial use, 30 per cent or nearly 10 million ha consists of eucalyptus, followed by acacias (3.9 million ha or about 12 per cent of the hardwood area) and teak (about 7 per cent). Short rotation plantations of hardwood species such as the eucalypts, acacias and *Gmelina arborea* have been grown for many years by the private sector, but the establishment of plantations of teak or other valuable hardwood species has been carried out only by government forest services because of their slow growth and hence delayed returns.

However, the likelihood of reduced supplies of high-quality hardwood logs derived from natural forests, combined with increasing purchasing power and expected higher prices for logs, is leading to increasing interest in investment by the private sector in valuable hardwood species, especially teak, in a number of countries, for example India, Malaysia, Costa Rica and Ghana. Of the softwoods grown for industrial purposes, fast-growing pines such as *Pinus radiata, P. patala and P. caribaea* constitute about 25 per cent of the area while other, often slow-growing pines make up about 36 per cent.

Responsibility for the monitoring and regulation of plantation crops grown for food and certain other purposes has long been the job of the agricultural sector. In recent years, however, several of these crops (the main ones being rubber, coconut and oil palm) have been providing 'forest' products that have been used for wood and fibre. For example, rubber wood is now used for the manufacture of about 80 per cent of the furniture made in Malaysia, while coconut and oil palm trunks and the branches of rubber wood are used for various forms of reconstituted 'wood'. Rubber wood and coconut stems are derived from the conversion of old plantations formerly disposed of by burning, while oil palm fruit residues are used for medium-density fibre board. The development of these new markets has thus not only improved financial returns but also used the resources in an environmentally friendly manner.

The area of these species appears to have increased from the 14 million ha reported in 1990. The increase may be due to better coverage of the data, although it is known that oil palm areas are increasing rapidly, rubber tree areas are also increasing and coconut plantations are decreasing. Coconut plantations comprise the largest area (about 42 per cent of the total), rubber 36 per cent and oil palm 22 per cent. Most of the coconut plantations are in Indonesia (33 per cent of the area) and the Philippines (28 per cent); most of the rubber plantations are in Indonesia (34 per cent), Thailand (20 per cent) and Malaysia (18 per cent), while most of the oil palm plantations are in Malaysia (44 per cent) and India (29 per cent). Not all of the areas mentioned above are suitable or available for substitute 'timber' production but they illustrate the potential.

Contributions of Forest Plantations to Wood Supply

The continuing increase in the area of forest plantations which have been established for industrial wood supply has been to meet the reduction of outturn foreseen from natural forests arising from deforestation and changes in land use (largely in the tropics and subtropics) or from natural forest being taken out of production and devoted to service functions such as conservation. It has been believed that the outputs from forest plantations can help to reduce the pressure on natural forests as sources of industrial wood supply; while the logging of tropical natural forests is not the prime cause of deforestation, logging roads often provide the means for farmers to gain access to forests. The reduction of logging, combined with effective protection, may thus help to reduce deforestation in certain locations until land use and ownership are clarified. However, none of these palliatives will remove the underlying causes of deforestation: high rates of population growth, poverty, hunger and a shortage of fertile land to cultivate.

The potential of forest plantations to meet demand for industrial roundwood is considerable; it has been estimated that the present global demand for paper pulp could be met from an area equivalent to only 1.5 per cent of the world's closed forest area. No global estimates of current output of timber from forest plantations are available, although FAO's global fibre supply model (GFSM) estimated that the potential annual growth of industrial wood from forest plantations in developing countries was about 5 per cent of the increment of natural forests in 1995. In some countries, plantation production already makes a highly significant contribution to the industrial wood supply, for example in New Zealand 99 per cent of industrial roundwood in 1997 was grown in plantations, while in Chile the equivalent figure was 95 per cent, in Brazil and Argentina 60 per cent, and in Zambia and Zimbabwe 50 per cent.

Estimating the future contribution of forest plantations to wood supply is at present imprecise and is based on many more or less unreliable assumptions, particularly concerning the rate at which afforestation will continue. By the year 2010 the GFSM (op. cit.) estimated that the potential increment from forest plantations would be about 40 per cent of that from natural forests in Asia, Oceania and Latin America and about 15 per cent in Africa, under rates of deforestation and afforestation largely the same as today.

FOREST LAND SURFACE

The world's forests cover an area of 3454 million ha or approximately 26.6 per cent of the land surface, with 56.8 per cent of this area in developing countries, which are mostly tropical. Forest cover has largely stabilized in most industrialized countries but deforestation continues in many developing countries. Between 1990 and 1995, the area of natural forests in developing countries decreased by an estimated 13.7 million ha per year, although this rate of loss appears to be slightly less than in the period 1980-90. Furthermore, forest management is relatively little practised at present in natural tropical forests.

The extent and condition of the global forest resource are determined by many economic, social and political factors external to the forestry sector, including, in particular, continued population growth and higher rates of global economic growth. Population growth, which will occur mainly in tropical developing countries, will continue to be combined with urbanization, while global economic growth will continue to be combined with changing consumption patterns, especially in the regions with fastest economic growth.

In the coming decades, pressures for increased food production are expected to lead to continued conversion of forest land to agriculture in many developing countries. It is estimated that in developing countries 90 million ha of land, of which about half may be forest land, would need to be converted to arable crop production alone between 1990 and 2010.

Agricultural land expansion is projected to be faster in sub-Saharan Africa than in the past; given the unsuit-ability of much of this zone for agriculture, there must be a continuing threat to natural forest cover from agricultural expansion as land continually goes out of agricultural production. Infrastructure development will also contribute significantly to the continued loss of forests. In addition to deforestation in developing countries, large areas of forest worldwide are being degraded by overharvesting, overgrazing, pests, disease, wildfires and airborne pollution.

At the same time as forest cover globally is decreasing and forests are being degraded, demands on forests to supply wood and non-wood products and social and environmental services are increasing. Global consumption of wood increased by 36 per cent between 1970 and 1994, and is expected to increase by another 20 per cent by 2010.

Greater emphasis is being put on the services that forests and trees can provide, including soil and water conservation, sequestration of carbon for the mitigation of climate change, conservation of biological diversity, support in combating desertification, enhancement of agricultural production systems, improvement of living conditions in urban and peri-urban areas, and provision of educational and recreational opportunities. Forests will remain an essential source of the livelihood of the poorer sectors of the world's population and a home to indigenous peoples for some time to come. The challenge of meeting the growing demand for forest products while safeguarding the ability of forests to provide a wide range of environmental services will increasingly be met through the planting of trees, either within the forest or outside. However, less natural or seminatural forest will be converted to plantations due to the emerging values of such ecosystems, and suitable land for plantation development will therefore be in short supply in many places. Governments will thus aim to encourage large- and small-scale landowners to plant trees outside forests, often integrated into agricultural systems, through policy measures including incentives.

There are likely to be moves towards making blocks of trees more 'natural' with a diversity of species, in some instances in order to provide a wider range of goods and thus serve as an insurance against the possibility of a single species failure or to guard against possible loss of soil fertility or site degradation, as well as to provide for improved amenity and recreational potential. There should be less emphasis on the area or quantity of forest cover and more attention paid to forest health and condition.

Forest cover in industrialized countries will continue to expand, and several newly industrialized countries will see their forest cover stabilize and even increase. Industrialized and newly industrialized countries will place greater emphasis on the conservation of natural forest where it still exists and on the conservation of seminatural forest. Plantation areas will continue to expand, either as intensively managed systems akin to farming practices or management will tend to move towards a more 'ecosystem' approach.

CONSERVATION AND MANAGEMENT OF BIOMASS RESOURCES MANAGEMENT

Conservation and management of biomass resources and environment are embedded in an array of knowledge and local practices. Most development projects tend to ignore the ingenuity of rural people in responding to the local problems based on their local knowledge. Initiatives intended to grow trees through Social Forestry projects have not taken cognisance of the existing local practises. For example in the Himalayan region, common practices of obtaining biomass resources from forests are through pollarding, pruning or selective cutting and not by felling of tress.

Similarly studies from Nepal have shown that, communities adjacent to forests mostly use dry wood as fuelwood and grasses for fodder that normally does not involve cutting of trees. For many local communities across South Asia, forests are not only of consumptive value, but also revered as places of great socio-cultural and religious significance. The other factors that regulate resource extraction from forests include (a) traditional rights (b) religious practices (c) and non-use of certain tree species in the region. For example, people have a strong belief that felling of certain tree species such as 'neem'

(*Azadirachta indica*) and 'pipal' (*Ficus religiosa*) brings evil to the members of the household. Thus local systems of knowledge and management are sometimes rooted deeply in their religion and belief systems, which help in conserving the resources. But, state management was often based on flawed premises or a lack of understanding of the people-nature dependency. Gradually local communities have lost access and control of the resources, resulting in conflicts with the state and other stakeholders.

In many parts of south Asia, local people maintained strict rules to manage natural forests in their vicinity even until the 60s. The examples given below are illustrative of the diversity of informal or traditional institutional arrangements at the local level:

(i) A grazing fee was levied on herders in some pasturelands. In some regions, grazing was banned in pond catchments to prevent siltation of ponds.

(ii) Cutting of grass and fodder leaves from the certain tree species, 'Sal' (*Shorea robusta*) in the *Terai* region in the Himalayas, or 'Palas' (*Butea monosprema.*) in central and western India was regulated on forestlands.

(iii) Collection of dry wood did not require individual permits from the informal councils at the village level, whereas special permits were required for felling trees on the hill slopes and catchments. In case of some species that are less in abundance such as Shisham (*Dalbergia sisoo*) and Bamboo (*Dendrocalamus* spp.) there was a ban on felling. These species had market value for their timber.

The village community performed collective work to enhance and maintain the tree cover, by fencing, planting, trenching in forests, fire lines maintenance and desilting of ponds. Each household had an obligation to contribute a share to such efforts, failing this its rights to resources were curtailed. These duties are no longer followed as the local rights to resources are curtailed with state control.

The indirect non-consumptive values were represented in day-to-day life in the form of rules and conventions regulating use of resources. The informal institutions tend to reduce the environmental uncertainty faced by resource users especially in arid and semi-arid regions and also contribute to the conservation of natural resources.

In some cases, the local institutions are better informed than the state departments about local ecological, economical and social conditions and about problems and constraints, which may be important from the management point of view. According to some studies, traditional management systems of CPRs have contributed to the protection of natural resources from being over used and have played a vital role in the management and conservation of natural resources.

TENURE RIGHTS AND BIOMASS RESOURCES

Prior to 1950s, local landlords controlled and had the authority to grant rights in forests in India, Sri Lanka and Nepal. In different areas, the tenure systems varied, for example the *Birta* system in mid hills of Nepal and the *Zamindari* system in northern part of India. The colonial regime in India and Sri Lanka significantly influenced the land tenure system. Under colonial management systems which continued in some areas even during the postcolonial period, the local landlords had the responsibility to manage forests and granted rights to the local households.

These include access rights, use rights and entitlement rights, which determine the access and end use of resources from the forests. Timber extraction in general was regulated, but there was free access to non-timber forest products. These resources were treated by local communities in many parts of south Asia as a "free good" which were easily available. But these rights and duties ceased with the land reforms during 50s and subsequent development policies initiated by respective governments. In Nepal, forests were nationalized as a part of a wider move to break feudal structures. In India, it was the Land Revenue Act in the respective states, which was implemented to nationalize forests. Nationalization of forests did more harm to landless and rural poor whose access to common pool resources was restricted as part of these resources were declared protected.

In addition, communal sources of fuel wood, fodder and non-timber products have been reduced as a result of the forestry projects under the management and control of state functionaries. Due to nationalization of forest, the lack of local tenure and rights meant there was no incentive for local users, especially the poor and landless to exercise restraint. The rights to access and use of forest resources was further restricted with social forestry interventions enclosing forestlands. For example, in the middle hill region of Nepal, the poorer sections of the people have less access to forest products for subsistence use and income today, than they had before the implementation of social forestry projects. This is also the case in India and Sri Lanka, where common lands, which were accessible to the landless prior to the social forestry intervention, are no longer freely available. In such a situation, the cultural overlay of rights creates paradoxically huge 'scarcity' differentials between households in villages, especially for the landless and poor. This tends to inhibit effective development and active participation by marginalized groups. Currently, the average poor household receives only one third to one fifth of the wood from forests than collected before the social forestry intervention. The rights to use resources from forestlands, which were formerly guaranteed by customary law, were banned after the land reforms and land acts came into existence in 1950s.

The choice of species or location of social forestry plantations was often made by people who had practically little local knowledge. The fact that there are local variations, even in small countries like Sri Lanka and Nepal, wherein women traditionally prefer certain tree species for fodder or fuel wood is important for any afforestation project to consider. The practice of use restrictions and the issue of land tenure, when viewed in the context of cultural and property rights based on gender inequalities and power relations within the households, presents a complex scenario. The tenure rights to land and resources in many South Asian regions are held by men, giving them legal rights over the resources, although women perform the frequent work of collecting biomass resources for household needs, cooking etc. Attached daily to the forest, women thus have a better knowledge on forest management and conservation, but are the most neglected group in social forestry projects.

There is no equity between men and women in land tenure specified officially in the land laws of the respective countries. Neither are women's rights in resource use are ensured in practice. Local traditions seems strongly influence on the women rights in land tenure, as it is the male child who inherits land. As a consequence, women's involvement in forestry projects is less, at the most they are being used as labourers in planting or watering the plants.

NUTRIENT CYCLING IN FORESTS

A nutrient cycle (or ecological recycling) is the movement and exchange of organic and inorganic matter back into the production of living matter. The process is regulated by food web pathways that decompose matter into mineral nutrients. Nutrient cycles occur within ecosystems. Ecosystems are interconnected systems where matter and energy flows and is exchanged as organisms feed, digest, and migrate about. Minerals and nutrients accumulate in varied densities and uneven configurations across the planet. Ecosystems recycle locally, converting mineral nutrients into the production of biomass, and on a larger scale they participate in a global system of inputs and outputs where matter is exchanged and transported through a larger system of biogeochemical cycles.

SOIL CHEMISTRY AND NUTRIENT VYCLING

By and large, the study of soil chemistry has concentrated on agricultural soils. The immediate practical questions to be answered have centred on the availability of plant nutrients, the need to develop chemical tests to define those soils suitable for particular crops and the prescription of the quantity of fertilizer that must be added to increase the productivity of the crop. With the exclusion of N (the availability of which is driven by biological processes in all soils), we might summarize the essential difference between the availability of plant nutrients in agricultural and forest soils:

- The availability of nutrients in agricultural soils is an immediate concern that mostly addresses inorganic equilibria over weeks or months;

- The availability of nutrients in forest soils is a long-term concern that should address biological processes and inorganic equilibria over years and centuries.

The growth of forests is a long-term process in which nutrients are cycled from plant to soil in litterfall and root turnover. Nutrients are withdrawn from tissues as they age and are translocated to actively growing tissues.

Timber harvesting takes away the aged tissues with relatively low nutrient concentrations, leaving the tissues with relatively high nutrient concentrations on site. In contrast, all of the nutrient supply for an annual crop is taken up from the soil within the few months the crop takes to reach maturity.

There is little cycling of nutrients from plant to soil. At maturity, the most nutrient-rich parts of the crop (seeds, leaves, storage organs) are harvested, leaving only the tissues with relatively low nutrient concentrations behind.

Thus, from the pioneering work of Ebermayer (1876) more than a century ago, much of the work on forest soils has been directed towards nutrient cycling, in particular:

- Quantifying the cycle of nutrients (uptake and return between plant and soil and retranslocation within the plant);
- Quantifying inputs and outputs, including nutrient removals in harvested timber.

While all of this is obvious, the science required is not obvious. Twenty years ago Stone (1979) concluded that despite a wealth of data on nutrient cycling processes for many forests, we do not have a sound experimental basis on which to assess the sustained productivity of forest ecosystems.

Ten years later, Landsberg *et al.* (1991) restated that conclusion: 'Conventional chemical analyses to determine the nutritional status of the soil yield information that may have little relevance in calculating the capacity of the soil to provide nutrients to trees, and the complexity of uptake processes through complex, dynamic and poorly defined root systems is tremendous.' Rather than attempting to review ongoing research on the experimental and analytical basis for assessing sustainability of nutrient supply, we give only a factual account of what is known of the chemistry and cycling of the major nutrients.

We do this in general terms, using data for a temperate forest of above-average productivity and we include a summary of the form, function, concentration and cycling of the major nutrients N, P, Ca, Mg and K.

Biomass

The above-ground mass of the forest (trees, under-storey and shrub layer) may reach 500t ha^{-1}. Litterfall *(A)* is 8tha^{-1} $year^{-1}$ but may reach 10tha^{-1} $year^{-1}$. If the litter layers (Q) weigh 30tha^{-1}, the decomposition rate of litter $k = A/(A + Q) = 0.21$ $year^{-1}$, and the half-life $t_{0\ 5}$ = 0.693/ft = 3.3 years. This and subsequent analyses are based only on the above-ground stand. Data for root mass and root turnover are both more limited and variable, and have been intensively summarized. We have previously estimated a rate of root turnover for our temperate forest of above-average productivity of 2-3tha^{-1} year^{-1}.

Nitrogen

The concentration of N in leaves is greater than the concentration of any other nutrient, but much of this N is withdrawn from leaves before they die and fall as litter.

Internal redistribution of N accounts for almost 30% of the total annual above-ground cycle. The half-life of N during litter decomposition is 3.5 years, a little longer than that for dry weight, so that N is immobilized during decomposition of the litter. The input of N in rainfall of 5 kg ha^{-1} year^{-1} is more or less balanced by the output of N in streamwater. More than 95% or so of N in the soil is in organic form; organic N is mineralized to NH_4^+-N and then under certain conditions to NO_3--N, a form freely available for uptake by plants. NH_4^+-N is held on negatively charged exchange sites of soils, but NO_3--N is mobile within the soil and is the form in which N moves with drainage waters into streams.

In soils of natural forests of the world, the generally high C/N ratio of litter and soil organic matter precludes mineralization proceeding to NO_3--N, so that the dominant form of N is NH_4^+-N. The cycling of N is therefore conservative, and N may be immobilized in litter of older forests (particularly in cooler-temperate areas) to the extent that the trees become N-deficient. Disturbance (treefall and the creation of gaps, fire, harvesting for timber) provides the conditions that result in a decrease in C/N ratio and the production of NO_3--N, in which form N may be leached from the soil into streamwater. Rapid regeneration of the forest, including rapid uptake of NO_3--N, following disturbance is therefore the key to the conservation of N.

Trace Elements (micronutrients)

The trace elements Fe, Mn, Cu, Zn, Mo and B have all played spectacular parts in agriculture and in tree plantations where plants have been introduced into new land. Applications of trace amounts (*e.g.* as little as 0.5kgha^{-1} of Mo) have corrected deficiencies that would otherwise have made agriculture or plantation forestry unproductive. We know of no such experience in natural forests, although there is little doubt that the distribution of many species may be determined by their ability to increase or restrict the supply of one or more of the trace elements. For example, concentrations of Mn in leaves and bark differ consistently and significantly between groups of the genus *Eucalyptus* but whether these differences are of significance in determining the distribution of species is unknown.

Phosphorus

Phosphorus occurs in highly mobile forms within plants and, like N, is withdrawn from plant parts during senescence; internal redistribution of P accounts for 50% of the total annual above-ground cycle. The half-life of P during decomposition is 4.1 years, *i.e.* P is immobilized throughout decomposition to a greater degree than N. A negligible amount of P comes in with the rain. P as H_2PO_4- is adsorbed by clay colloids and is then highly immobile in the soils so that losses of P to streamwater are negligible.

However, in forest soils more than 50% of the total P in surface soils may be in organic form. Mineralization of organic P is therefore fundamental to ecosystem function, and P availability is 'determined by competition between biological and geochemical sinks for phosphate anions'. Whereas methods for estimating rates of N mineralization have been developed and are in general application, there are no methods for routinely estimating the rate of P mineralization.

The difficulty is that whenever H_2PO_4- comes into solution, there is a surface to adsorb it. This competitiveness between biological and chemical sinks for P remains a priority area of research in quantifying the long-term sustainability of P supply for forests.

Calcium, Magnesium and Potassium

Ca, Mg and K, and the non-essential element Na, are generally grouped as the major exchangeable cations in soil. An exchangeable cation is held adsorbed on a negatively charged surface of a colloid (both inorganic and organic). In this exchangeable form, the cations are not soluble but are brought quickly into solution and made available for uptake by exchange of protons from plant roots.

Ca is highly immobile within the plant, where it is incorporated within cell walls. Thus 80% of the total annual above-ground cycle of Ca is in lit-terfall. In contrast, K is highly mobile and K^+ easily leached from tree crowns by rain; leaching accounts for about 30% of the total annual above-ground cycle of K.

Roots and the Rhizosphere

Roots supply plants with water and nutrients and provide anchorage in soil. Most studies of roots in forests have been directed to the surface horizons, where fine roots (< 2 mm in diameter) are concentrated and where water and the bulk of nutrients are sourced. There are few studies of roots below 1 m, with only nine reported in the review of Jackson *et al.* (1996). In general our knowledge of root characteristics, including biomass, distribution and fine-root area, is poorly developed compared with the above-ground components of forests. Much of the recent interest in root distribution and turnover in forests has arisen primarily from concern for increased C flux from soils due to climate change.

Distribution, Mass and Turnover of Roots

A global analysis of root distribution in terrestrial biomes showed that boreal forests have the shallowest rooting profiles (80-90% of roots in the upper 30 cm of forest soils) and that temperate coniferous forests have the deepest rooting profiles (52% of roots in the upper 30 cm). When considered together, all temperate and tropical trees have 26% of roots in the top 10 cm and 60% in the top 30 cm.

Root biomass of forests has been estimated at 2-5 kg m^{-2}, with the highest values in tropical evergreen forests. However, discrepancies between model estimates and measured characteristics of fine-root length and surface area point to the need for more data on root biomass and distribution, and for more careful measurement of fine root biomass in particular.

These analyses of root distributions and characteristics (biomass, surface area and nutrient content) among forests will enable improved modelling of water and nutrient uptake on a global scale. They also provide the basis for improving global models of the impact of climate change on vegetation distribution and C sequestration in soil. When linked with models that predict the change in vegetation distribution and type, they will allow better prediction of the consequences of global environmental change. The need for this type of analysis seems all the more urgent when it is considered that roots can account for more than 50% of net primary productivity (NPP) in forests and that the global pool of C in fine roots is 5% of the size of the atmospheric C pool. Because about half the global pool of C in fine roots is in forests, further study of factors influencing fine-root turnover is critical to the understanding of CO_2 flux between terrestrial and atmospheric C pools.

While roots are concentrated in the upper metre of soil in most forests, deep roots become important in forests subject to periodic drought because they provide access to water held deep in the soil. Carbon from these deep roots may contribute over decades to net C release from soil following conversion of forest to pasture, as shown by Nepstad *et al.* (1994) in an Amazonian rain forest. These sources of deep soil C have not been generally considered in estimates of C flux from forest soils, especially those of tropical regions where forests continue to be cleared for agriculture.

The turnover of fine roots has been investigated in relation to forest disturbance, including clearcutting, in the Northern Hardwood ecosystem. Biomass of fine roots (<2mm) in mature hardwood forest was 471 gm^{-2}, with average lifespan ranging from 8 to 10 months. Biomass of fine roots recovered to 71% of that in the mature forest within 3 years after clearcutting.

Rhizosphere

The rhizosphere is the area of soil immediately surrounding the root where biotic and abiotic processes interact to create an environment distinct from bulk soil further from the root. Compared with bulk soil, rhizosphere soil is usually more intensively weathered, has a lower pH and has greater concentrations of cations and P. Exudates released from plant roots and microorganisms benefit plant growth by increasing nutrient acquisition and metal detoxification, by alleviation of anaerobic stress in roots and by their action in mineral weathering.

There is no doubt that some forest species can modify the rhizosphere to access previously unavailable reserves of soil P by releasing organic acids that solubilize inorganic P. For example, Grierson (1992) demonstrated that in P-deficient soils, *Banksia integrifolia* induced the development of short-branched, tertiary, lateral roots (proteoid or cluster roots) that released citric and malic acids (the main organic acids induced under P deficiency). Organic acids probably increase the availability of P in soils through a combination of decreased adsorption of P, increased solubiliza-tion of P compounds and chelation of metals such as Fe.

Other rhizosphere processes that impact on the availability of soil P reserves include increase in root hair length and density, release of C that enhances mycorrhizal exploitation of bulk soil, and the release of phosphatases that solubilize organically bound P.

The role and function of mycorrhizas in forest soils has been reviewed by Brundrett (1991) and Vogt *et al.* (1991) and there is no doubt that mycorrhizal associations benefit trees by enhancing nutrient acquisition, especially in relation to P. The evidence for N and P transfer to forest trees via mycorrhizal associations has been demonstrated in tracer studies run over relatively short time-courses. However, it is only recently that new techniques have allowed longer-term studies; these have confirmed the benefits of mycorrhizas to nutrient acquisition and have also demonstrated increased plant growth as a result. Further work of this kind is required to establish the relative contribution of mycorrhizal uptake mechanisms to nutrient acquisition in mature forests over the long term.

The activity of microorganisms in the rhizos-phere is driven by the release of low-molecular-weight organic compounds by the root and there is some evidence to suggest that the growth of certain microorganisms is favoured while that of others is suppressed. For example, Gonzalez *et al.* (1995) reported that while microbial numbers were greater in the rhizosphere of *Alnus,* roots differentially favoured colonization by proteolytic and ammonifying organisms and inhibited nitrifying organisms.

Norton and Firestone (1996) reported a greater than 50% increase in N turnover (mineralization and immobilization) in the rhizosphere relative to bulk soils in *Pinus ponderosa* microcosms. Further evidence of the complex interactions in the rhizosphere come from reports of ectomycorrhizal fungi stimulating bacterial growth associated with the release of citric acid by the fungus and, conversely, of inhi bition of bacterial growth. Recent work has aimed at characterizing the chemical and biotic nature of the rhizospere, especially as it is influenced by soil liming and nutrient addition.

Most work has been focused in the vicinity of industrialized areas of the Northern Hemisphere, where there are concerns for the unimpeded functioning of beneficial root-rhizosphere reactions under both controlled (fertilizer inputs) and uncontrolled (atmospheric deposition) inputs of acidity, N and S. Reported effects of ammonium sulphate application on rhizosphere chemistry vary and many of the changes, such as increased acidity and Al, are found in bulk soils as well.

Other effects include depletion of P and K from the rhizosphere of Norway spruce in southern Sweden induced by ammonium deposition and the subsequent stimulation of P and K uptake.

The benefits of rhizosphere organisms, including mycorrhizal fungi, in maintaining the growth of tree seedlings in nurseries has long been known. This concept has been extended recently to the development of soil inoculation techniques to improve the growth of seedlings in the reforestation of degraded or clearcut sites, where inoculation with forest soil has increased the number of ectomycorrhizas.

Differential effects on survival and growth of seedlings, where planting holes were inoculated with soil transferred from forestry or plantation sites, have been reported. Increases in seedling growth have been associated with increased nutrient mineralization brought about by soil animals stimulating microbial turnover in transferred soil, while increases in survival probably resulted from the introduction of beneficial rhizosphere organisms.

Another dimension in the use of mycorrhizas to benefit tree growth has been the recognition by Garbaye (1994) of bacteria that selectively promote mycorrhizal development (the so-called 'mycorrhization helper' bacteria, MHB). In anticipation of the applications discussed by Chanway (1997) the process of selecting and using MHB cultures for controlled 'mycorrhization' has been patented by Garbaye and Duponnis (1991).

VILLAGERS SUPPORTED LOCAL MANAGEMENT OF FOREST

The management and use of forest resources and forest outputs seldom falls neatly into private, state or common property categories. Many, if not most, forests are characterized by multiple products and multiple users and are usually being managed in pursuit of multiple objectives. They therefore often exhibit a variety of overlapping tenure arrangements. There may be individual, corporate and collective rights to use of a resource that forms part of state property. Several different groups of users may have tenure rights to different products or to use at different times. Use may be managed or take place in an uncontrolled open access situation.

The villagers supported local management of forest. Lahru who could provide the leadership was able to challenge the traditional leader and was encouraged to expand his constitution to cover development issues. While, the onset of JFM, should have helped Lahru retain this leadership, villagers instead shifted their loyalty to the young leaders who were supported by a more responsive NGO. While, JFM makes it easier for the villagers to secure forest patches to meet their requirements, they clearly favour a situation where this security can be enlarged to other areas of rural life.

The Forestry Case of West Bengal

Bhagwati Chowk village, in Gopegarh Forest Beat in Midnapore East Forest Division, has a near uniform population make up, and is next to a relatively larger patch of forest. Poltu Singh has been the leader and spokesperson for the successful FPC since inception. Various features typical for Bengal mark the area.

Many of the villagers live on land they got through left government's land re-distribution schemes. Poltu Singh has been a leader with *Panchayat* and much of his confidence comes from his familiarity with bureaucracy. However, currently the *panchayat* is under the schedule list and the council leader is a woman from schedule caste from the neighbouring village of Amratoli. Poltu Singh, however, remains a leader of his village FPC and his 'patch' of Gopegarh forest.

The FPC, however, is less radical than Ober in Bihar. A 'successful' FPC in terms of good forest cover and with several instances of harvest and usufruct distribution, the committee is, in general, disinterested in forest activities on a day to day basis.

It is active only during the official get together with Forest Department officials and infrequent committee meetings. The FPC has, however, created an 'exclusion zone' (similar to Ober) in the forest that lie under their jurisdiction, and takes up occasional fights with infiltrators from neighbouring village, often with villagers from same *panchayat*. It has often served as a platform to work in forestry micro-plans (when funds were available from the Forest Department) and with NGOs The impact that NGOs have had on rural development forestry is beyond the scope of this chapter. It is, nonetheless, worth mentioning that an NGO that could provide support in various areas of rural development were accepted more by the community than those who, for example, worked only to promote JFM activities.

There are, however, zones of anomaly in the region. The neighbouring villages of Amratoli and Phulpahari form another FPC and share a large block of forest. While Amratoli is in same *Panchayat* as Bhagwati Chowk, Phulpahari belongs to another *panchayat*. The population make-up is diverse and so are the forestry practices and dependency. The caste-Hindu settlements are better off than the rest of the population, and their women folk do not go to forest for collection of fuelwood. The political factions are apparent, and forestry is often compromised for political gains. The entire forest, for example, was cut down during the *panchayat* election in 1998.

Existing Social Capital

The norms of reciprocity and attitude towards trust are dynamic in nature, and choices are made in such a manner as to maximize the benefits to individuals. Once the common goals are within sight, the community builds on the existing social capital and makes adjustments to capitalize on the incentives. It seizes upon the new opportunities and does not mind risk breaking established conventions.

It also indicates the regulatory constraints of the system. A successful and popular village level forum cannot overspecialize in one issue such as forestry alone. A rigid dichotomization cannot be held, for example, between forestry and other rural development issues.

Also, a populist agenda, as is common with new organizations such as NGO-club in Ober, may, in order to attract membership, harm the common pool resources like forests. Left to them, village institutions would, while seizing on the most attractive incentive, tend to compromise on old obligations, thus making the system inherently unstable. While in case of Bihar, on an NGO's involvement the villagers switch loyalty to capitalize on the opportunities for development, in Bengal *panchayat* leaders barter away forest timber in exchange for votes in an election.

What is, however, unmistakable is that despite a marginal nature of FPCs (as compared to the established FD, *panchayats*, NGOs), these Forest Protection Committees often function to create a power-base and legitimacy of their own. The legitimacy that the FPC commands in Bhagwati Chowk is, for example, unambiguous. Poltu Singh despite his not being a council leader remains the leader in the eyes of Gopegarh villagers because of his formal control over a vital resource.

Informal Niche in JFM Arrangement

The section deals with informal In an effort to locate the informal practices, it is important to note that these activities are termed informal only when seen against the norms and rules that are laid out in the Joint Forest Management agreement. Many practices are not only essential to villagers' day to day's needs but have been in practice (noticed or disguised, contested or informally recognised) for a long period of time. Activities that find support in JFM's participatory atmosphere.

Through time, the Indian State has exercised a near total control over its forest-land and forest citizens. The current participatory paradigm of sharing revenue and forest-usufruct through Joint Forest Management places this control (and polity) in an ambiguous position. The section argues that while the state still maintains a strong hold over both tenure and forest land practices, the new arrangement has made inroads for unprecedented concessions that allow rural people greater control over the village and protected (community) forests.

PROTECTION AND MANAGEMENT OF 'PROTECTED' FOREST

The FPCs, as a part of the JFM agreement, in return for protection and management of 'protected' forest would, apart from access to fuelwood, share a percentage of cash with the state on sale of timber. A range of concessions allowing collection of fuelwood and minor forest produce have existed (pre-dating the current participatory forestry) in both the States, though the relation between the Forest Department and citizens has been one of constant confrontation and suspicion.

Bihar, traditionally, has enjoyed more rights over forest use than Bengal. There is no fee payable, in Bihar, for fuelwood collection or livestock-grazing, for example.

The tribal population has also been entitled to a share of timber on occasional harvests. There is, however, no provision (including the JFM scenario) in either of the states to allow felling of trees other than those announced through government-circulars (sanctioned through official 'working plans') and carried out at the initiative of the Forest Department. The use of forest, nevertheless, is not limited to what is prescribed within the JFM agreement; villagers regularly extract logs of wood from protected forest for construction of houses and agricultural tools.

The forest bureaucracy is now no longer seen as a foe against whom village communities must unite against, often turning forest land into 'open access resources'. Instead, villagers are identifying more solidly with isolated forest patches that lie under the jurisdiction of village Forest Protection Committees. Forest lands are now demarcated and governed, for management purposes (protection, regeneration, harvesting, even silvicultural decisions), by these committees. The committees often decide on the exclusion of neighbouring village-communities from use of 'their' forests. To 'catch and punish the erring population' frequently forms the basis of protection of forests and to establish norms onto villagers from outside and even within the community itself.

CHANGES IN THE FOREST NATURAL ENVIRONMENT

Forests and woodlands are an important part of our landscape and provide many benefits to society. The tree species that are native to the UK have adapted to the local climate, atmosphere and soils over many years. However, human activities have resulted in changes to the natural environment, especially over the past 200 years. It is expected that the climate of the UK will become milder and wetter in winter, and significantly hotter and drier in the summer months over the coming century. These changes to our climate are predicted to be larger and more rapid than any since the last ice-age, posing real problems for trees, woodland and forestry.

Climate changes directly and indirectly affect the growth and productivity of forests: directly due to changes in atmospheric carbon dioxide and climate and indirectly through complex interactions in forest ecosystems. Climate also affects the frequency and severity of many forest disturbances.

In the context of climate change, sustainably managed forests – and the products derived from them – play an essential mitigating role. Forests are one of the globe's greatest carbon-sequestration tools, and sustainable forestry naturally creates an endless cycle of carbon absorption and storage.

Trees and forest products play a critical role in helping to tackle climate change and reduce greenhouse gases. As trees grow, they clean the air we breathe by absorbing carbon dioxide from the atmosphere, storing the carbon in their wood, roots, leaves or needles and surrounding soil, and releasing the oxygen back into the atmosphere. Young, vigorously growing trees absorb the most carbon dioxide, with the rate slowing as they reach maturity.

When trees start to decay, or when forests succumb to wildfire, insects or disease, the stored carbon is released back into the atmosphere. In any of these cases, the carbon cycle begins again as the forest is regenerated, either naturally or by planting, and young seedlings once again begin absorbing carbon. Manufacturing wood into products requires far less energy than other materials – and very little fossil fuel energy. Most of the energy that is used comes from converting residual bark and sawdust to electrical and thermal energy, adding to wood's light carbon footprint.

Climate is a strong influence on forestlands in B.C. It affects tree growth, productivity, and numerous resources derived from these lands. By maintaining biodiversity in our forests, we can help ecosystems to withstand environmental changes such as climate change.

STRENGTHEN ADAPTIVE CAPACITY OF FORESTS

Negatively affect forests and many of their plant and animal species. In addition, they may negatively affect the availability of other resources, necessary for species survival. Current forest composition and structure are however, the result of past changes in climate and shows that forests and their species have an inherent capacity to adapt to change.

The main differences of current climate change with historic changes are the increased rate of these changes and the degraded and fragmented state of the remaining forests, which reduces the capacity of the species and ecosystems to adapt. The challenge is to help species and ecosystems to adapt to climate change while at the same time ensuring that ecosystem services are maintained.

This will require the identification of the changes to which the forest will need to adapt.

Locally, changes may be disastrous, unless climate, ecosystem and species changes are accompanied by adjustments in the local social and economic systems. For example, increased occurrence of severe fires will require greater collective action to prevent fires as well as improved weather and fire danger forecast services . Companies producing furniture of high value species from natural forests, whose natural regeneration under changed climate conditions has become increasingly difficult, may have to change geographic range for their inputs, or change to other species and/or other processing procedures. Communities and private landowners depending on local forests may have to change livelihoods after severe hurricane damage.

Nationally or at the landscape level, changes may be slower and less disastrous in the short term. New challenges include the identification of those species groups and ecological processes that are essential for the most important ecosystem services. This would include in most cases identification of water catchment areas (hydrogeology) and the role of forests in maintaining water quality and quantity. It will be important to increase the probability that changing ecosystems will continue to provide the important services and goods. In particular, ecosystems in geographic locations at the extreme limits of climatically well-defined areas, such as mountainous forests, rangelands and boreal forests, are likely to be severely affected and may disappear. Some authors suggest that maintaining functional diversity and 6 composition will preserve ecosystem services, while others found that different functional groups will react differently to environmental changes, indicating that climate change may favour some functional groups over others. More research is needed however, to identify those functional groups essential for the desired ecosystem services and goods in particular areas and to understand how these can be conserved and protected.

Reduce Risk and Intensity of Pest, Disease and Fire Outbreaks

Reducing the climate induced risk of pests, diseases and fire outbreaks, in particular, in dry areas and less diverse forests will be a major environmental challenge. Breeding of more resistant or more resilient varieties is a medium to long-term solution for plantation species, although, that introduces new risks because strengthening the adaptive capacity of a species for one trait may weaken it to other traits. Identifying species for their "realized fitness" - for example, varieties of a species that survived insect attacks, diseases or fires, similar to the expected events in a particular region - and then facilitating their migration to the area of interest, may be another strategy. In both cases, identification of the traits that will increase resistance or resilience will be important as will be replicating those traits over generations and successfully introducing the species or varieties in the area of interest, without introducing new problems (such as undesired invasion).

Predicting future changes in pest and disease outbreaks and adjusting management accordingly is another option, which requires the development and validation of models that reliably predict impacts under different climate and management scenarios. A further option is the identification and implementation of forest management systems that are known or thought to reduce the risks of pests, diseases and/or fires.

While there are several well known means to protect forests and plantations, in many cases these are not applied for a variety of reasons, or are not applied to those forests most in need. The challenges are to identify and address the reasons for the lack of application of management techniques and to adjust management options to the threats in a participatory, socially and economically acceptable manner.

CHANGES IN SOCIOECONOMIC ENVIRONMENT

Risk of Migration into Forest Areas

Climate change will affect all people but in particular, rural people that depend on nature for their livelihoods, and poverty stricken communities in the urban-rural interface that are often subjected to the consequences of extreme weather events. Climate change is expected to change the aptitude of lands for specific crops, cause problems of droughts, fire and flooding and may drive many people from their lands. These people are likely to either go to cities to look for jobs, often adding to urban poverty, or to other rural areas to look for other lands where they may be able to continue their agricultural livelihoods or find employment in the agricultural sector. The surge of interest in fuels from biomass (*e.g.* corn, sugarcane and oil palm) adds another dimension to this migration. The purchasing of land, often based on speculation, in the hope of selling later for higher prices to investors interested in biofuel production, may cause migration. The expected high incomes from biofuels may also motivate landowners to convert their forests into energy plantations, oftentimes in an unsustainable manner. On the other hand, if well planned, biofuels could also help avoid or reduce migration by providing off-farm employment.

Forest use values, even in the case of the most successful enterprises, will not be able to compete with oil palm or other energy crops in those lands suitable for the crops. Legal definition of user and owner rights of forest areas and the mechanisms to defend those rights will be important elements of strategies to prevent unauthorized entrance into forests. Market mechanisms that restrict trade of products from companies that do not show social and environmental responsibility in their production and purchase policies may be another strategy. An individual forest user or owner will find it difficult to influence legislation, their implementation or the way that markets function. Collaboration with other stakeholders, neighbours, value chain members, and state administrators will be essential to the development of adequate measures to reduce the conversion and degradation of forests. Forest users and owners, however, have a longstanding tradition of independence and in the past have not shown tendencies to such collaboration. Lack of trust (often justified), has often hampered relations between different stakeholders in the forest and environmental sectors. Building sufficient trust to facilitate collaboration may be the biggest challenge of all for future forest management and needs the collaboration of all actors involved.

8

Joint Forest Management for Forest Conservation

CONSERVATION ISSUES IN THE HIMALAYAN REGION

The Himalayan region is bestowed with varied landscape features that provide multitude of habitats to a diverse array of faunal communities including several species of wild ungulates. The region covers nearly 11% of India's geographical area and range from sub-tropical to alpine zones. The region is well known for extensive alpine biome, temperate conifer and broadleaf forests, sub-tropical (foot-hill) forests, temperate grassy slopes, and various other habitats.

Despite tremendous potential, most of the areas in this region exhibit low abundance of wild ungulates. It is an irony that there are more Himalayan Tahr (Hemitragus jemlahicus) in New Zealand in comparison to the Himalayan region. This fact generally speaks of the status of wild ungulates in the region. Most of the ungulates are found in small isolated pockets and largely restricted to Protected Areas (PAs). Even within the PAs, especially in the eastern Himalaya and adjacent hill states, wildlife and ungulate densities are very low. It is not surprising that a naturalist, having conducted large mammal surveys in the eastern Himalaya, is known to have commented, "the stage is beautiful but many of the actors are missing".

Questions on mountain ungulate conservation in the Greater Himalaya range from 'What are the reasons for low ungulate abundance in the Himalayan region' to 'What are the major issues pertaining to their conservation' and 'What are the ways and means to achieve better conservation status for mammals in general and the ungulates in particular'. We address these issues and strategies for conservation for mountain ungulates in the Greater Himalaya in this article.

MAJOR CONSERVATION ISSUES

Several factors have led to low abundance and poor conservation status of ungulates in the Himalayan region. Firstly, the Himalayan ecosystem is relatively young, fragile and low in primary productivity. With the increase in human population the area has undergone rapid degradation, fragmentation and loss of wildlife habitat.

Poaching and trade of animal parts, competitive exclusion by the domestic livestock, faulty land use practices, and human-wildlife conflicts are other factors affecting wild ungulates and other faunal communities. Some of the issues related to conservation of wildlife ungulates.

Habitat Degradation and Fragmentation

(i) Alpine Habitats of the Greater Himalaya: The region above natural tree line (ca. 3300-3600 m above mean sea level in the western and north-western and ca. 3600-3800 m in the central and eastern Himalaya) represent alpine habitats which are characterized by treeless vegetation, alpine scrub and meadows. The ungulates frequently inhabiting this zone are Blue sheep (Pseudois nayaur), Himalayan musk deer (Moschus chrysogaster), and Himalayan tahr. The major causes leading to degradation and fragmentation of alpine habitats include overgrazing by livestock, commercial harvest of wild medicinal herbs, uncontrolled tourism and mountaineering in certain areas. The area is very prone to soil erosion, avalanches and landslides owing to steep and fragile terrain. Very few PAs in the region give complete protection to alpine habitats except a few (*e.g.*, Valley of Flowers NP and Nanda Devi National Park [NP] in Uttaranchal). However, a majority of the PAs remain neglected in terms of management even though they represent important habitats for typical faunal communities.

(ii) Sub-alpine Forests: The area between ca. 3000 m and natural 'tree line' represents an important ecological belt throughout the Himalaya. Besides Himalayan musk deer and serow (Nemorhaedus sumatraensis), it forms summer habitat for two important ungulates viz., Hangul (Cervus elaphus hanglu) in the west (Kashmir Valley) and Takin (Budorcas taxicolor) in the east (Mishmi hills, Arunachal Pradesh). Of all the habitats in the Himalayan region, the sub-alpine forests have undergone maximum degradation and fragmentation owing to anthropogenic activities such as collection of non-timber forest produce (including montane bamboo, mushroom, medicinal and aromatic plants), poaching, livestock grazing and camping by the herders (Awasthi et. al. communicated). Sathyakumar *et al.* (1993) have reported that increased livestock grazing and associated impacts have led to low musk deer densities in many areas in Kedarnath Wildlife Sanctuary. The subalpine forests, 'tree line' and the alpine scrub interspersed with alpine meadows is the optimal habitat for musk deer but this habitat, particularly the 'tree line' has degraded in many parts of the Himalaya due to cumulative impacts of livestock grazing. Most graziers prefer to camp and graze their livestock in and around 'tree line' due to availability of fuel wood, water, food and cover for livestock.

(iii) Montane forests of North-Western and Western Himalaya: The forested habitats in middle elevation rages (1500-3000 m) in the western and north-western Himalaya exhibit a great diversity of flora and fauna. The major vegetation types include Himalayan Dry Temperate (conifer), Himalayan Moist Temperate (broadleaf), and several other categories. In many areas (especially on south facing gentle slopes) the forests have been transformed into the scrub jungle and cultivation. The common ungulates of the forested habitat include Himalayan musk deer, serow, goral, sambar (Cervus unicolor), barking deer (Munitacus muntjac) and wild pig (Sus scrofa). Much of the forested habitats in the region are affected by encroachment for habitation and cultivation, livestock grazing, lopping of trees for fodder. These activities have led to failure of regeneration and resultant change in the structure and composition of forests. Rawat *et al.* (1999) have reported that plant species diversity have changed with an increase in unpalatable plant species in the fringes of Kedarnath Wildlife Sanctuary as a result of human use. The forests of Western Himalaya are rich in wild mushrooms including highly prized morel (Morchella esculenta). Local people, in large groups, visit several parts of such PAs in order to collect this mushroom thereby causing heavy disturbances and affecting the threatened and sensitive species of fauna such as Himalayan musk deer and pheasants.

(iv) Central and East Himalayan Montane Forests: Major forest types in the middle elevation (1500-3000 m) ranges of Sikkim (Central Himalaya) and Arunachal Pradesh include Temperate Broadleaf Forests, Conifer forests and bamboo brakes. These forests support Himalayan musk deer, serow, takin, several other ungulates such as Asian elephant (Elephas maximus), gaur (Bos gaurus), sambar, wild pig and various other mammals including primates, and carnivores. But for the occasional fire (localized in certain areas) and slash and burn agriculture in some parts of Arunachal Pradesh, much of the forested habitats in the central and eastern Himalaya are intact.

(v) The Shivaliks and sub-Himalayan Forests: This zone (<1500 m) represents the sub-tropical climate, varied topography, rich alluvial soils and intermingling of taxa from the Indo-malayan and Palaearctic regions. The major forest formations, according Champion & Seth (1968) include Sub-tropical Dry Evergreen Forests, Sub-tropical Pine

Forests, Northern Dry Mixed Deciduous Forest, Dry Shivalik Sal Forest, Moist Mixed Deciduous Forest, Sub-tropical Broadleaf Wet Hill Forest, Northern Tropical Semi-evergreen Forest, and Northern Tropical Wet-evergreen Forest. The Shivalik hills are best represented between the Ganges and Yamuna rivers in Uttaranchal. The entire belt all along covers an area of ca. 40,000 km^2 of which only <2100 km^2 area falls under PAs represented by Simbalbara WS, Rajaji and Corbett NPs. Ecologically, entire Shivalik belt is considered as highly sensitive zone. This region suffers heavy fragmentation and degradation of habitat due to human encroachment and proliferation of exotic weeds such as Lantana camara, Parthenium hysterophorus, Cassia tora, and Sida spp. Hajra (2002), based on the analysis of remote sensing data of Shivalik zone in Uttaranchal found that even though south facing slopes appear to be suitable for goral, but only about 13% area was moderately suitable and less than 1% area is highly suitable for this species. Rest of the area is either forested or very steep and not available for goral.

(vi) The North-Eastern Hills: The natural landscapes and habitats in this region have been extensively modified due to shifting cultivation or slash and burn (Jhum) agriculture. Besides, pressure on the land due to exploitation of forest for timber and lack of a scientific forest management has led to proliferation of exotic weeds and degradation of forests. With tremendous increase in human population during recent decades the jhum cycle has come down from 20-30 years to about 5 years and even up to 3 years in many areas. Reduction in jhum cycle due to increase in pressure on land has accelerated the process of habitat degradation and fragmentation which has affected ungulates and other faunal groups in the region.

(vii) Temperate Grassy Slopes: The temperate belt in the western and north-western Himalaya support an extensive grassland habitat which is largely anthropogenic in nature. These grassy slopes have developed largely on the south facing, steeper slopes which cannot be cultivated and are burned during winter to promote grass growth. Such slopes are grazed and maintained as hay slopes or 'Ghasnis' in many sectors of Western Himalaya. The steeper and inaccessible areas are occupied by Himalayan tahr and goral. Goral is one of the prominent species of mountain ungulates that has evolved in this habitat in the region (Mishra 1993, Pendharkar 1993). The gentler slopes close to human habitation have degraded over the years and goral habitat has been reduced considerably. The temperate grassy slopes and adjacent woodlands of Dachigam NP, Kashmir valley support a highly threatened subspecies, the Hangul or Kashmir stag.

The grassy slopes of Dachigam are also reported to have degraded considerably over the years (Khursheed Ahmed, personal communication).

Habitat Loss

Excessive degradation and fragmentation eventually leads to habitat loss. Systematic studies documenting loss of ungulate habitats in the Himalayan region are lacking. Hence, it is difficult to point out clear cases of this phenomenon. However, in most of the sectors, there are plenty of evidences indicating the shrinkage of wildlife / ungulate habitat. Green (1986) reported that over 70% of potential musk deer habitat has already been lost due to habitat loss and habitat degradation in the southern side of the Greater Himalaya.

Based on the spatial time-series analysis of remote sensing data, Awasthi (2001) has pointed out that in upper regions of Bhagirathi Valley, Garhwal Himalaya there has been a considerable increase in the area of human inhabitation and cultivation during last 30 years. This study also indicates that much of the sub-alpine and temperate broadleaf forests have converted into scrub vegetation. Conversion of forested habitat into scrub, is in a way loss of habitat especially for the sensitive ungulates such as Himalayan musk deer. Displacement of human populations for the developmental projects, construction of roads along the sensitive habitats, encroachment of forests for agriculture and heavy infestation of exotic weeds are other causes of habitat loss in the region.

Competition with Domestic Livestock

Owing to high seasonality and low primary productivity, the Himalayan region supports relatively low ungulate / herbivore biomass. It is therefore, obvious that with the increase in the biomass of domestic livestock in many areas, wild ungulates have suffered competitive exclusion.

Rawat (1998) has pointed out that several areas in the Himalaya there is an overstocking of livestock leading to decreased productivity and degradation of pastures. Sathyakumar *et al.* (1993) have reported that increased livestock grazing and associated impacts have led to low musk deer densities in many areas in Kedarnath Wildlife Sanctuary. Although animal husbandry is one of the main stay of livelihood in the Himalayan region, the management of livestock especially disease surveillance, rotational grazing and pasture management have been neglected leading to conflicts with wildlife as well as PA managers.

Poaching: All mountain ungulates of the Greater Himalaya are seriously threatened due to poaching for meat, skin/hide, and for products such as the 'musk' from Musk deer.

Poaching for meat (bush meat hunting) is common in many parts of the Greater Himalaya, particularly in the Eastern Himalaya. Species such as the goral, tahr, serow, takin are poached for meat/hide by local villagers and indigenous people throughout their distribution range. There are instances of wild mountain ungulate meat served in local restaurants in towns or villages adjoining wilderness areas. Although, the extent of poaching by local villagers or indigenous people is not known, there are evidences of the consequences of high poaching levels in many areas where mountain ungulates have become either locally extinct or occur in very low densities.

Poaching for sport and meat by the personnel of the security forces in the international border areas, also have led to serious impacts on mountain ungulate populations.

Poaching of musk deer for 'musk' is rampant through out the Greater Himalaya for its high commercial value (about US $ 65,000/kg in 1985) in the international markets (Green 1986, Sathyakumar 1993b). This has led to local extinctions of this species in many parts of the Greater Himalaya. As a result, the once continuous distribution of musk deer is now confined to some isolated pockets and in most of these areas they occur in very low densities. Similarly, poaching of Himalayan tahr and goral have either resulted in local extinctions or very low densities in many areas.

CONSERVATION ISSUES IN THE TRANS-HIMALAYA

A variety of physical, biotic and political characteristics of the Trans-Himalaya influence conservation issues that are peculiar to the region. Chief among the challenges is the limited resources available to the native population of the region. These populations which mostly occur at low to moderate densities of <2 persons per km2, are primarily agro-pastoral, or, as in the Tibetan Plateau zone, are largely nomadic pastoralists. Human populations are increasing in the region with the breakup of the traditional polyandrous system and with fewer people opting for becoming celibate monks and nuns. An important factor that needs to be considered is that in the harsh Trans-Himalayan landscape there is hardly any area that is not already in use by people at some time or the other during the year. Arable land is mostly limited to alluvial fans, and some stable areas in the valley bottoms.

Almost all the available arable land is already under cultivation and all pastures are grazed by the domestic stocks at least seasonally. Addition of newer areas under some poverty alleviation schemes incorporate development of expensive and long flow irrigation systems. Our observations show that these most frequently end up as failures since most of such channels are damaged by avalanches or are washed away by floods or simply break up due to the unstable substrate. Thus in spite of numerous efforts any significant addition of arable land is remote. National parks and sanctuaries in India do not permit consumptive use and require resettlement of people outside such areas. The point that we are driving at is that the Trans-Himalayan region has very specific and limited area for cultivation or use as pastures and thus offers no or very few alternatives for resettlement of people outside PAs.

The region has other peculiarities such as very poor road access, power supply, and difficult communication, apart from a harsh climate during much of the year. These features of the region thus do not allow scope for conventional industrial development and employment in the region, as is possible in other regions of the country.

As far as the wildlife values are concerned a very important characteristic of the Trans-Himalayan area is that it provides almost continuous wildlife habitat. Almost the entire landscape has large mammals, including the snow leopard and wolf, but the densities may vary greatly from very poor areas to small pockets that may be rich in some large mammals. This means that a large amount of wildlife may actually be occurring outside existing PAs. In Nepal, for example, over 60% of the snow leopards are thought to occur outside PAs. In India, our coarse estimate is that of the probable maximum of 600 snow leopards in India, ca. 80% may be occurring outside PAs. Among other endangered wildlife species such as the Tibetan antelope, Tibetan gazelle, Tibetan argali, brown bear, and kiang, the entire or substantial populations occur outside existing PAs. Mishra, (2001). has found that in Spiti, the Tabo area that is not within any PA has bharal densities that are higher than the Kibber Wildlife Sanctuary.

Shah (1996) has found remnant populations of the highly endangered Tibetan gazelle and argali in unprotected areas in northern Sikkim. The paradox here is that in spite of a very large proportion of the Trans Himalaya being under the PA network, numerous endangered species continue to occur outside. With this, and the fact that the region in general is resource deficient, with few livelihood options for the local herders we need to reconsider the approach of having large national parks and wildlife sanctuaries as 'inviolate areas' in the region. This is a challenging situation and the lack of a vision for conservation, as usually defined by Management Plans for PAs is absent at present.

Conservation issues common to the area primarily relate to deficiencies of infrastructure and staff for PA management, grazing competition between wild and domestic herbivores, conflicts relating to damage to crops and livestock by wildlife, some levels of poaching of snow leopard and prey species, wildlife diseases and political issues. These issues were also flagged as the most important ones in a meeting between scientists and managers at Leh and of numerous snow leopard experts from all over the snow leopard range countries. We now examine these issues in some detail:

Infrastructure and Staff for PA Management

In the harsh environs of the region, there is a severe shortage of staff, effective infrastructure and funds for the management of the PAs. For example Ladakh has ca. 13,100 km^2 under its PA network with a total park staff of merely about 20. This translates to 655 km^2 to every park staff-a completely ineffective strength to manage the region under any circumstance, but especially so under the difficult climatic and topograhic conditions in the Trans-Himalaya.

The numerous existing tasks of the Wildlife Department range from protection, tourism management, verification of compensation claims to nature education activities in the PAs spread all over the ca. 45,000 km^2 Ladakh region. We also understand that most park staff lack the necessary clothing, equipment and housing necessary for effective work in the region. The situation in Lahul & Spiti is not very different. It is an urgent requirement for the Centre and State Governments to provision the necessary resources to these areas for effective conservation in the region.

FOREST SOIL NATURAL RESOURCES AND CONSERVATION SERVICE

The Natural Resources Conservation Service (NRCS), formerly known as the Soil Conservation Service (SCS), is an agency of the United States Department of Agriculture (USDA) that provides technical assistance to farmers and other private landowners and managers.

Its name was changed in 1994 during the Presidency of Bill Clinton to reflect its broader mission. It is a relatively small agency, currently comprising about 12,000 employees. Its mission is to improve, protect, and conserve natural resources on private lands through a cooperative partnership with local and state agencies. While its primary focus has beenagricultural lands, it has made many technical contributions to soil surveying, classification and water qualityimprovement. One example is the Conservation Effects Assessment Project (CEAP), set up to quantify the benefits of agricultural conservation efforts promoted and supported by programmes in the Farm Security and Rural Investment Act of 2002 (2002 Farm Bill). NRCS is the leading agency in this project.

PROGRAMMES AND SERVICES

As of November 2011 the NRCS consists of 42 programmes and activities. The NRCS also offers services to private land owners, conservation districts, tribes, and other types of organizations.

Farm Bill

The conservation provisions in the Food, Conservation, and Energy act of 2008. This bill provides conservation opportunities for farmers. Services also include financial assistance which makes sure money is allocated and used properly in providing conservation for various natural resources, technical assistance which is given through the Conservation Technical Assistance programme (CTA). This service is available to anyone interested in conservation of natural resources and easements. This includes:

Farm and Ranch Land Protection Programme

(FRPP) The purpose of this programme is to work with land owners to purchase development rights to current farm and ranch land, in order to keep said land from being developed for other uses. The programme matches funds from property owners, and can be applied whether potential buyers are private or from state or local government, as well as Native American tribes. In order to receive funds from the programme, the land must be privately owned, and have an offer for sale pending. The land must be large enough to support substantial agricultural yield, and be surrounded by land with a similar nature. If there is a danger for soil erosion, a conservation plan must be included.

Grasslands Reserve Programme

(GRP) Volunteer programme to increase animal and plant biodiversity, and to protect grasslands. Participants limit use of grassland for commercial and agricultural development. The land may still be grazed or seeded, with the exception of the nesting seasons of bird species that are protected under law. A grazing management plan must be submitted for participation.

Healthy Forests Reserve Programme

(HFRP) Landowners volunteer to restore and protect forests in 30 or 10 year contracts. This programme hands assisting funds to participants.

The objectives of HFRP are to:

- Promote the recovery of endangered and threatened species under the Endangered Species Act (ESA)
- Improve plant and animal biodiversity
- Enhance carbon sequestration.

Wetlands Reserve Programme

(WRP) Volunteer programme for landowners to protect or restore wetlands on properties they own. The programme offers both financial and technological support to these landowners in order to help cultivate long term wetland health with optimal biodiversity per acre.

NRCS National Ag Water Management Team

(AGWAM) Serves 10 states in the Midwest United States in helping to reduce Nitrate levels in soil due to run-off from fertilized farmland. The project began in 2010 and initially focused on the Mississippi Basin area. The main goal of the project is to implement better methods of managing water drainage from agricultural uses, in place of letting the water drain naturally as it had done in the past. In October 2011, the The National "Managing Water, Harvesting Results" Summit was held to promote the drainage techniques used in hopes of people adopting them nationwide.

Snow Survey and Water Supply Forecasting

Includes water supply forecasts, reservoirs, and the Surface Water Supply Index (SWSI) for Alaska and other Western states. NRCS agents collect data from snowpack and mountain sites to predict spring run-off and summer streamflow amounts. These predictions are used in decision making for agriculture, wildlife management, construction and development, and several other areas. These predictions are available within the first 5 days of each month from January to June.

Wildlife Habitat Incentive Programme

(WHIP) Is a volunteer programme for improving habitats for wildlife on farmlands, private lands, and Indian land. The programme was renewed in 2008 under the Food, Conservation, and Energy Act. WHIP provides up to 75 per cent cost share and technical assistance, and includes both terrestrial and freshwater aquatic habitats. The programme aims to protect seven threatened species in particular:

- Lesser Prairie Chicken
- New England Cottontail
- Southwestern Willow Flycatcher
- Greater Sage-Grouse
- Gopher Tortoise
- Bog Turtle
- Golden-winged Warbler

Conservation Technical Assistance Programme

(CTA) Is a blanket programme which involves conservation efforts on soil and water conservation, as well as management of agricultural wastes, erosion, and general longterm sustainability.

NRCS and related agencies work with landowners, communities, or developers to protect the environment. Also serve to guide people to comply with acts such as the Highly Erodible Land, Wetland (Swampbuster), and Conservation Compliance Provisions acts. The CTA can also cover projects by state, local, and federal governments.

Gulf of Mexico Initiative

(GoMI)Is a programme to assist gulf bordering states (Alabama, Florida, Louisiana, Mississippi, and Texas) improve water quality and use sustainable methods of farming, fishing, and other industry. The programme will deliver up to 50 million dollars over 2011-2013 to apply these sustainable methods, as well as wildlife habitat management systems that do not hinder agricultural productivity, and prevent future over use of water resources to protect native endangered species.

International programmes

The NRCS (formerly SCS)has been involved in soil and other conservation issues internationally since the 1930s. The main bulk of international programmes focused on preventing soil erosion by sharing techniques known to the United States with other areas. NRCS sends staff to countries worldwide to conferences to improve knowledge of soil conservation.

There is also international technical assistance programmes similar to programmes implemented in the United States. There are long term technical assistance programmes in effect with one or more NRCS staff residing in the country for a minimum of one year. There are currently long term assistance programmes on every continent. Short term technical assistance is also available on a two week basis.

These programmes are to encourage local landowners and organizations to participate in the conservation of natural resources on their land, and lastly landscape planning has a goal to solve problems dealing with natural resource conservation with the help of the community in order to reach a desired future outcome.

TECHNICAL RESOURCES

Water

Pollution of water due to a number of different pollutants has driven the NRCS to take action. Not only do they offer financial assistance but they also provide the equipment needed for private land owners to protect our water resources. Water gets polluted by nitrogen and phosphorus which causes algae to grow proliferously causing the oxygen concentrations to decline rapidly, life is no longer supported in this habitat. Excessive sedimentation is also another concern along with pathogens threats that can find their way into water systems and cause detrimental effects. NRCS works in a way to help both the land owner and the water systems that need prevention or restoration.

Water Management

Water management strives to manage and control the flow of water in a way that is efficient while causing the least amount of damage to life and property.

This helps provide protection in high risk areas from flooding. Irrigation water management is the most efficient way to use and recycle water resources for land owners and farmers. Drainage management is the manipulation of sub surface drainage networks in order to properly disperse the water to the correct geographical areas. The NRCS engineering division is constantly making improvements to irrigation systems in a way that incorporates every aspect of water restoration.

Water Quality

A team of highly trained experts on every aspect of water is employed by the NRCS to analyze water from different sources. They work in many areas such as: hydrology and hydraulics, stream restoration, wetlands, agriculture, agronomy, animal waste management, pest control, salinity, irrigation, and nutrients in water.

Watershed Programme

Under watershed programmes the NRCS works with states, local governments, and tribes by providing funding and resources in order to help restore and also benefit from the programmes. They would like to provide: watershed protection, flood mitigation, water quality improvement, soil erosion reduction, irrigation, sediment control, fish and wildlife enhancement, wetland and wetland function creation and restoration, groundwater recharge, easements, wetland and floodplain conservation easements, hydropower, watershed dam rehabilitation.

Plants and Animals

Plants and animals play a huge role in the health of our ecosystems. A delicate balance exists between relationships of plants and animals. If an animal is introduced to an ecosystem that is not native to the region that it could destroy plants or animals that should not have to protect itself from this particular threat. As well as if a plant ends up in a specific area where it should not be it could have adverse effects on the wildlife that try to eat it. NRCS protects the plants and animals because they provide us with food, materials for shelter, fuel to keep us warm, and air to breathe. Without functioning ecosystems we would have none of the things mentioned above. NRCS provides guidance to assist conservationists and landowners with enhancing plant and animal populations as well as helping them deal with invasive species.

Fish and Wildlife

NRCS for years has been working towards restoration, creation, enhancement, and maintenance for aquatic life on the nearly 70 per cent of land that is privately owned in order to keep the habitats and wildlife protected. NRCS with a science based approach, provides equipment to wildlife and fish management. They also do this for landowners who qualify to benefit from these technologies.

Insects and Pollinators

Pollination via insects plays a huge role in the production of food crop and flowering plants. Without pollinators searching for nectar and pollen for food the plants would not produce a seed that will create another plant. NRCS sees the importance of this process so they are taking measures to increase the declining number of pollinators. There are many resources provided from the NRCS that will help any individual do their part in conservation of these important insects. Such as Backyard Conservation which tells an individual exactly how to help by just creating a small habitat in minutes There are many others such as: Plants for pollinators, pollinators habitat in pastures, pollinator value of NRCS plant releases in conservation planting, plant materials publications relating to insects and pollinators, PLANTS database: NRCS pollinator documents. All of these are valuable resources that any individual can take advantage of.

Invasive Species and Pests

Many adverse effects are present due to invasive species. Plants and animals both inhabit areas that they are not intended to be. The kudzu vine for example covers miles of foliage. These invasive species cause America's reduction in economic productivity and ecological decline. Humans are unknowingly transporting these invasive species via ships, planes, boats, and their own bodies. NRCS works in collaboration with the plant materials centers scattered throughout the country in order to get a handle on the invasive species of plants.These centers scout out the plants and take measures to control and eradicate them from the particular area.

Invasive animals such as feral hog, european gypsy moth, and the sirex woodwasp pose a significant threat to America's wildlife as well as to the health of human beings. The hog was introduced as a food source for humans, but now the swine pandemic is a serious threat to humans. They gypsy moth destroys natural forests that are habitat to many beneficial species. The Woodwasp feeds on pine trees as well as providing a means of transportation for a fungus that kills pine trees.

Livestock

Livestock management is an area of interest for the NRCS because if not maintained valuable resources such as food, wools, and leather would not be available. The proper maintenance of livestock can also improve soil and water resources by providing a waste management system so that run off and erosion is not a problem. The NRCS provides financial assistance to land owners with grazing land and range land that is used by livestock in order to control the run off of waste into fresh water systems and prevent soil erosion.

Plants

Plants are a huge benefit to the health of ecosystems. NRCS offers significant amounts of resources to individuals interested in conserving plants. From databases full of information to financial assistance the NRCS works hard to provide the means needed to do so. The plant materials programme, Plant materials centers, Plant materials specialists, PLANTS database, National Plant Data Team (NPDT) are all used together to keep our ecosystems as healthy as possible. This includes getting rid of unwanted species and building up species that have been killed off that are beneficial to the environment. The NRCS utilizes a very wide range of interdisciplinary resources.

The NRCS also utilizes the following disciplines in order to maximize efficiency:

- Agronomy
- Erosion
- Air Quality and Atmospheric Change
- Animal Feeding Operations and Confined Animal Feeding Operations
- Biology
- Conservation Innovation Grants
- Conservation Practices
- Cultural Resources
- Economics
- Energy
- Engineering
- Environmental Compliance
- Field Office Technical Guide
- Forestry,
- Maps
- Data and Analysis
- Nutrient Management
- Pest Management
- Range and Pasture
- Social Sciences
- Soils, and Water Resources

These Science-Based technologies are all used together in order to provide the best conservation of natural resources possible.

FOREST, SOIL AND WATER-CONSERVATION MEASURES

The key to soil and water conservation is the utilization and treatment of land according to its capability.

LAND-CAPABILITY CLASSIFICATION

Any soil and water-conservation project includes two distinct sets of operations, viz.

1. The mapping of land for classification according to its capability,
2. Planning and executing measures to check erosion, improve land productivity and reclaim wasteland.

The farm plans for effective soil and water conservation are based largely on the capability of the land. The land-capability classification map is normally prepared by interpreting a standard soil-survey map. Land-capability classification is a systematic arrangement of different kinds of lands according to those properties that determine the ability of the land to produce crops on a virtually permanent basis.

The Factors Determining Land-capability

These are the major soil characteristics of the land, *e.g.,*, the texture of the top soil, its effective depth, permeability of the top soil and subsoil, and associated land features, *e.g.,*, the slope of the the land, the extent of erosion, the degree of wetness and susceptibility to overflowing and flooding. The grouping of soils into capability classes is done primarily on the basis of their capability to produce common cultivated crops and pasture plants without deterioration over a long period.

Land-capability Classes

The land-capability classes are based on the intensity of hazards and the limitations of use.

The land-capability classes range from the best and most easily farmed land to that which has no value for cultivation, grazing or forestry, but which may be suited to wild-life, recreation or for watershed protection.

They all fall into 2 broad groups: one suitable for cultivation and other land uses, and the other not suitable for cultivation, but suitable for other land uses.

Land Suitable for Cultivation and Other Uses

Class I (Green Colour)

Soils in class I have very few or no limitations that restrict their use.

This type of land is nearly level and the erosion hazard is low. The soils are deep, well-drained, easily worked, hold water well and are either fairly well supplied with plant nutrients or highly responsive to the application of fertilizers. The soils are not subject to damage because of overflow.

The local climate must be favourable for growing many of the common field crops.

In irrigated areas, the soils may be in class I, if the limitation of the arid climate has been removed by relatively permanent irrigation works. These soils need ordinary management practices to maintain productivity. Such practices may include the use of one or more of the following: fertilizers, lime, cover and green-manure crops, conservation of crop residues and crop rotations. Soils in this class are suited to a wide range of plants, may be used for cultivated crops, pastures, forests, and wildlife, food and cover.

Class II (Yellow Colour)

Soils in class II have some limitations which reduce the choice of plants or require simple conservation practices.

The limitations of soils in class II may result from the effects of one or more of the following factors:

- A gentle slope,
- A slight susceptibility to erosion,
- Less than ideal soil depth,
- Occasional damaging overflow,
- Wetness which can be corrected by drainage, but existing permanently as a moderate limitation,
- Slight to moderate salinity or sodium, easily corrected but likely to recur,
- A slight climatic limitation on soil use and management.

These soils require careful management. The limitations are only a few and the practices are easy to apply. They may need one or more of the following practices: terracing, strip cropping, contour cultivation, water disposal area, covered with vegetation crop rotation, cover and green-manure crops, stubble mulching, the use of fertilizers, manure and lime. These soils may be used for growing cultivated crops, raising pastures, forests, and for wild-life, food and cover.

Class III (Red Colour)

Soils in class III have moderate limitations which reduce the choice of plants or require special conservation practices. Soils in class III have more restrictions than those in class II and, when used for cultivated crops, the conservation practices are usually more difficult to apply and to maintain.

Limitations of soils in class III may result from the effects of one or more of the following factors:

- A moderately sloping land,
- Moderately susceptibility to water or wind erosion,
- Frequent overflow accompanied with some crop damage,
- Very slow permeability of the sub-soil,
- Wetness or continuing water-logging after drainage,
- Shallow soil depth up to the bed-rock, hard-pan or clay-pan which limits the rooting-zone and water storage,
- Low moisture-holding capacity,
- Moderate salinity or sodium,
- Moderate climatic limitations.

Class IV (Blue Colour)

Soils in class IV have severe limitations that restrict the choice of plants and require careful management. The restrictions in the use of these soils are greater than those in class III and the choice of plants is more limited. When these soils are cultivated, very careful management is required and the conservation practices are more difficult to apply and to maintain.

The use of these soils for cultivated crops is limited as a result of the effect of one or more of the permanent features, such as:

- Steep slopes,
- Severe susceptibility to water or wind erosion,
- Severe effect of past erosion,
- Shallow soil,
- Low moisture-holding capacity,
- Frequent overflow accompanied with severe crop damage,
- Excessive wetness or continuing hazard of water-logging after drainage,
- Severe salinity or sodium,
- Moderately adverse climate.

These soils can be used for crops, pastures, forests, and wild-life food and cover.

Land Not Suitable for Cultivation but Suitable for Other Land Uses

Class V (Dark Green or Uncoloured)

Soils in class V have little or no erosion hazard, but have other limitations, the removal of which is not practicable. They are used largely for pastures, forests, and wild-life food and cover. Such land is nearly level and is not subject to more than slight wind or water erosion. Cultivation is not feasible because of one or more limitations, such as overflow, stoniness, wetness or severe climate.

Examples of class V land are:

- Soils of lowlands subject to frequent overflows which prevent the normal production of cultivated crops,
- Nearly level soils with growing season that prevents the normal production of cultivated crops,
- The level or nearly level stony or rocky soils,
- Ponded areas where drainage for cultivated crops is not feasible but where soils are suitable for grasses or trees.

Soils in class V are not suitable for raising cultivated crops, but are suitable for perennial vegetation (grazing and forestry, with few or no limitation). Pastures can be improved, and benefits from proper management can be expected. Physical conditions of soils are such that it is practicable to apply pasture improvements, if needed, such as seeding, liming, fertilizing, and water control with contour furrows, drainage ditches, diversions of water spreaders.

Class VI (Orange Colour)

Soils in class VI have severe limitations that make them unsuitable for cultivation and limit their use largely for pastures, or forests, or wild-life food and cover. Soils in class VI have continuing limitations which cannot be corrected, such as:

- Steep slope,
- Very severe erosion hazard,
- Very severe effect of past erosion
- Stoniness,
- Shallow rooting-zone,
- Excessive wetness or overflow,
- Low moisture capacity,
- Salinity or sodium,
- Severe climate.

Soils in this class are subject to moderate limitations under grazing or forestry use.

Class VII (Brown Colour)

Soils in class VII have very severe limitations that make them unsuitable for cultivation and restrict their use largely to grazing, or forestation, or wild-life food and cover. The soils in this class are subject to severe limitations or hazards under either grazing or forestry use. The physical condition of soils is such that it is not practicable to adopt pasture improvements and water-control practices. Soil restrictions are sever e than those in the case of class VI soils.

Class VIII (Purple Colour)

Soils and land forms in class VIII have limitations that preclude their use for commercial plant production and restrict their use to recreation, wild-life food and cover or to water-supply, water shed protection or for aesthetic purposes. Significant return on site benefits from soils and land forms in class VIII cannot be expected from management of crops, grasses or trees, although indirect benefits from wild-life, watershed protection or recreation may be possible.

Limitations which cannot be corrected may result from the effects of one or more of the following factors:

- Erosion or erosion hazard,
- Severe climate,
- Wet soil,
- Stones,
- Low moisture capacity,
- Salinity or sodium.

Bad lands, rock outcrops, sandy beaches, marshes, deserts, river wash, mine tailings and other nearly barren lands are included in class VIII in order to protect other more valuable soils to control water or for wild-life or for aesthetic reasons. The land-capability class is indicated on the maps by roman numerals I to VIII or by standard colours or by both.

TECHNOLOGY OF WATER CONSERVATION

Water conservation technologies cover all methods of conserving water through increasing water use efficiency, enhancing capacity to retain runoff water, and eliminating water pollution. Water use efficiency largely depends on availability and adoption of water saving devices and willingness of the consumers to reduce their total water consumption volumes. Furthermore, existing rules and regulations such as pricing mechanisms, reduce the total volume used, while economic incentives largely affect the choice and adoption of technology for water conservation technologies.

Water conservation processes can broadly be categorized into pre consumer and post consumer based approaches. Pre consumer based approaches involve increasing the efficiency of water extraction, storage and conveyance. Usually a large amount of water is lost either through evapotranspiration or seepage during transfer from the abstraction point to the point of use. It is estimated that, of the total rainfall in Thailand, about 70% returns to the atmosphere through the process of evaporation and transpiration. Using technologies that can minimize losses such as reducing evaporative losses from reservoirs, seepage losses from canals and water application losses prior to the water being used for economic purposes, can conserve a vast amount of water. In contrast, post consumer based approaches include the use of marginal quality waters, such as slightly saline water from the sea, untreated groundwater from shallow tubewells and rainwater collected from thatched roofs, for washing, toilet flushing purposes, etc.

In Asia, the agricultural sector consumes more than 75% of the total water withdrawn from all sources. About 60% of this water is lost during conveyance and distribution. Engineering technologies to reduce these losses, and to enhance utilization of water in irrigation schemes, are well studied and established practices which are beyond the scope of this book. In contrast, agronomic technologies such as efficient and modern methods of irrigation (including drip irrigation, sprinkler irrigation and surge irrigation) are less well-known and may be considered water conservation technologies. These technologies are documented in standard textbooks on agricultural and water resources engineering, and are briefly reviewed herein.

Recycling is also an alternative, post consumer technology for conserving and augmenting water supplies. It involves the reuse of water previously used for one purpose for a particular use in another application, before it reaches a natural waterway or aquifer. By using water several times, farms, urban areas and industries can increase the productivity of each litre of water consumed.

Industries can conserve water by changing production processes from open to closed systems. In many industrial plants it is possible to recycle the cooling water. Some industries process water several times, and treat it at the end of its period of usefulness, prior to discharging the water to a natural water course. Reuse conserves raw water and, at the same time, reduces the volume of wastewater as well as wastewater treatment costs substantially. Several case studies on water conservation practices in India are described. In terms of particular technologies, there are no specific technologies involved, but rather a way of managing conservation practices in individual households and industries. Some of the technologies typically being practised in the Asian Region also are presented herein.

DUAL WATER DISTRIBUTION SYSTEM

Technical Description

Water reuse, and the reuse of wastewater in particular, is receiving increasingly wide attention, even though it is often considered to be of marginal quality. Use of treated wastewater through dual water distribution and plumbing systems can provide a secondary source of water for purposes such as irrigating private gardens and toilet flushing. The provision of waters of lesser quality through a separate distribution system for non-potable purposes from alternative sources of supply can help lower the demand for potable freshwater. Application of this technology is largely a matter of cost, acceptance and practice, as the distribution technology involved is not significantly different in dual distribution systems compared to conventional single distribution systems.

In the Kathmandu Valley, where water scarcity is increasing, conjunctive use of shallow groundwater sources for toilet flushing and washing of clothes, together with potable water supplied through the municipal water supply system for other purposes, is being practised in residential areas. A rower pump (hand pump) is fitted to a 3.8 cm diameter polyvinylchloride pipe ranging from 6 m to 15 m in length which is driven to the ground. Water from the groundwater source thus tapped is pumped manually whenever needed.

Extent of Use

This technology is widely used in the Kathmandu Valley in locations where the groundwater table is within 15 m of the ground surface. It is also popular in locations where the municipal water supply is intermittent. People of low income are increasingly using this technology in preference to the higher cost municipal supply.

Operation and Maintenance

The operation and maintenance of dual distribution systems is simple, and involves keeping the pipe and the pump clean. Maintenance consists of changing the pump washer once a year or whenever it starts leaking. No additional maintenance of the municipal supply system is required, and no changes in municipal distribution system operation are necessary.

Level of Involvement

Providing dual distribution systems at the household level requires no external involvement relative to the groundwater sourced portion of the system. The municipal sourced portion of the system is generally constructed and operated by the local governmental unit.

Costs

The total cost of the groundwater sourced portion of the system is about $55 for a 10 m deep well.

Effectiveness of the Technology

This technology has helped alleviate the problem of water scarcity in Kathmandu. For an average household, more than 60% of the annual household water requirements is met by using shallow groundwater, which is of lower quality than the municipal water, for non potable purposes.

Suitability

Water supplied from the rower pump can be used for toilets, bathing, gardening, car washing, and similar purposes. The dual distribution system is most suitable for use in areas where the groundwater is within 15 m of the ground surface; otherwise, a mechanical pump will be necessary, adding to the cost of the groundwater sourced portion of the system.

Advantages

This technology is inexpensive and can be constructed using locally-available technology. Water is made available whenever needed, and use of the dual sourced system eases the water scarcity problem not only at the household level but also throughout the entire city.

Disadvantages

Conjunctive use of dual sourced water may be limited as a result of poor water quality. Water drawn from alternative sources may not be used for drinking even after boiling because of odours and tastes associated with groundwater. Further, such water may pose severe health hazards if the abstraction point is located too close to septic tank outflows.

The possible high nitrate concentrations and bacterial levels that may be present in surfacial groundwaters may also lead to health hazards when used by poor people and children for drinking purposes. Widespread use may contribute to a decline in the groundwater table (due to over exploitation). Also, water logging and mosquito breeding in the pump area may occur if proper drainage is not provided.

Cultural Acceptability

No cultural problems have been noted, although use may be limited due to odours associated with the groundwater.

Further Development of the Technology

This technology will be more attractive and useful if simple and inexpensive electric motors are attached to the tubewell and the groundwater sourced portion of the system is fully incorporated into the household water supply system.

EVAPORATION REDUCTION

Water Evaporation Retardants

In India, the use of Water Evaporation Retardants (WER) is being investigated as a water conservation measure in surface waters. Control of evaporative losses is being effected using Ceto-Stearyl alcohol, a blend of saturated fatty alcohols, previously imported at high cost and not readily available in India. India's first fatty alcohol plant was commissioned in 1981 at Jalgaon, Maharashtra, by Aegis Chemical Industries Ltd., using an exclusive technology based on an high pressure hydrogenation technology developed in collaboration with Haldor Topsoe, Denmark.

The plant is one of the few in the world and the only one in India, producing alcohols to international specifications. In 1983, Aegis successfully developed an effective water evaporation retardant, Acilol TA 1618 WER, an emulsion based on fatty alcohols manufactured from natural vegetable oils. Acilol TA 1618 WER was developed in the laboratory after extensive research aimed at identifying a product compatible with the climatic conditions of the country. Field trials were conducted in March 1984 with the help of Gujarat Engineering Research Institute. These trials confirmed the effectiveness of Acilol TA 1618 WER. A dispensing technique for applying Acilol TA 1618 WER to water surfaces, consisting of barrel tanks with a drip feed arrangement mounted on "Floating Rafts" anchored at different points on the lakes or reservoirs, was also developed to suit Indian conditions. The first major project using this technique to conserve water was undertaken by the Public Health Engineering Department (PHED), Jaipur, at Ramgarh Lake. About 50 mg/m2/d of the chemical is required and can result in a savings of about 30% of the daily loss of water due to evaporation.

Technical Description

Evaporative losses can be controlled using various technologies. For example, the use of mono molecular organic surface films has been shown to be an efficient technology for reducing such losses. The mono molecular film is applied to an open surface water storage area and allowed to over the water surface.

Typical mono molecular films are comprised of long-chain fatty alcohols such as cetyl alcohol (hexadecanol) and stearyl alcohol (octadecanol). These chemical not only suppress evaporation but also prevent mosquito breeding in the water.

The use of stearyl alcohol in doses of up to 70 g/cm2 can reduce evaporative losses by up to 55% of the loss due to evaporation from a free water surface. However, the economics and extent of use of this technology are yet to be explored and established.

Extent of Use

Use of surface films to reduce evaporative losses is principally in arid and semi-arid regions.

Operation and Maintenance

The operation and maintenance requirements of this technology are negligible. Maintenance is required only for the rafts and boats, which may be locally constructed, used during the application of the evaporation retardants. No specialized skill is required.

Level of Involvement

The implementation of this technology is generally carried out by government departments, primarily the public health engineering and water supply departments.

In India, it is now being promoted by various state government agencies and local authorities.

Generally, control of evaporation is focused at the government level having prime responsibility for water resources management.

Costs

No cost data were available as this technology remains largely experimental.

Effectiveness of the Technology

Of the various substances capable of forming mono molecular layers on a water surface, fatty alcohols in their pure form have been found to be most suitable and effective in retarding evaporation with no side effects.

Savings from the prevention of water loss due to evaporation have been reported to be as high as 0.70 million cubic metres of water (equivalent to one month's water supply for Jaipur City, India) using Cetyl and Stearyl Alcohol as an evaporation retardant.

Suitability

The technology is most suitable in small surface water storages where there are no strong winds to disrupt the retardant layer.

Advantages

Use of evaporation control techniques requires a small capital investment. Locally constructed rafts can be used in its application, and skilled labour is not required, except for operation of motorized boats.

Fatty alcohols present no hazards in handling, and are non-inflammable, non-toxic, and non-irritating. There are no known harmful effects, making this technology safe for application on drinking water lakes and reservoirs. There are also no known ecological ramifications, as the alcohols contain straight chain carbon compounds which are biodegradable and permeable to oxygen.

Disadvantages

The fatty alcohols used as WERs are not readily available and are costly. Historically, fatty alcohols were produced only by sperm whales or by sodium reduction of animal oils and fats. It was only after the development of high pressure hydrogenation process that good quality fatty alcohols are now commercially available. However, this is a very advanced technology, and requires trained and skilled staff to operate.

Cultural Acceptability

There are no known problems with cultural acceptability.

Further Development of the Technology

More research is necessary in the area of stabilising the chemical film in the face of high wind velocities. Properties such as rate of spreading, specific resistance to evaporation and surface viscosity are required to be measured in evaluating the efficiency of the retardant.

COCONUT PICK-UPS

Technical Description

In South India, coconut pick-ups is a name popularly used for weirs constructed exclusively to provide water to coconut gardens. These weirs or pick-ups are small structures built across the seasonal or perennial streams to slow the flow of water at an appropriate location. This results in surface water storage, groundwater recharge, reduction of soil erosion and availability of water for other purposes.

The bunds that form the puck-ips are constructed of locally available materials such as stones, boulders or mud turfed with grass. The bunds are usually built within the water course almost to the height of the surrounding ground level, depending upon the width and steepness of the stream.

When stream flows occur, the coconut plantation, situated within the floodplain on either side of the water course, is temporarily flooded. The water impounded by the structures recedes within 3 to 4 days, with the excess water frequently diverted into a tank or cistern. Sometimes there will be several coconut pick-ups within a catchment, and several tanks which may flow from one tank to the other.

Extent of Use

This technology has been used in many places in South India.

Operation and Maintenance

The pick-up is constructed by the farmers and hence both operation and maintenance is the responsibility of the user. The principle maintenance requirement is keeping the pick-up sealed to prevent loss due to leakage.

Level of Involvement

Usually, the pick-ups are managed by the users who build them. Repair and maintenance are also carried out by the user. Government may be involved in the funding of the pick-ups, and in the initial permitting of the project site. This function is commonly performed by the local self-governing body that functions as a farmers/users association and manage their activities.

Costs

Construction of a typical coconut pick-up for an area of 10 ha in India would cost about $2 000.

Effectiveness of the Technology

The technology is successful in providing adequate water to many coconut farmers for agricultural purposes.

Suitability

The technology is suitable for use in tropical areas with moderately rolling topography and small streams.

Advantages

The use of pick-ups encourages groundwater recharge by promoting infiltration and vertical percolation. This recharge helps to sustain yields to percolation tanks and tube wells supplying open tanks. The recharge of open wells is almost immediate, even at a distance of up to 0.5 km depending on soil structure and gradients. Ponds created by the pick-ups can serve also as a drinking water source for livestock from up to 10 villages.

Further, the floodwaters deposit a few millimetres of silt behind the pick-up structure and enrich the soils of the floodplains behind the pick-up with nutrients associated with materials like manure, leaves and other terrestrial debris carried in the runoff flows. These minerals and nutrients enrich the soil particularly during the monsoon season without the use of fertilizer supplements.

Disadvantages

The presence of the temporary ponds behind the pick-ups can promote mosquito growth and exacerbate human health risks.

9

Biodiversity Assessment in Joint Forest Management

BIODIVERSITY

In recognition of the significance of its biodiversity assets, Guyana signed the UN Convention on Biological Diversity during the Earth Summit of 1992. The Convention commits signatories to adopt regulations to conserve their biological resources. In this context, as stated in the NEAP: "the Government is committed to saving biodiversity, studying biodiversity and using biodiversity sustainably and equitably. The Government is also aware of the need to ensure that the ownership of intellectual property rights, including knowledge and customary and traditional practices of local people, is adequately and effectively protected. However, to date, Guyana lacks specific legislation to facilitate such actions."

Although our wealth in biological diversity is unquestionable, there is a general paucity of information on its exact nature and extent. However, some 8,000 species of flora, of which half are endemic, have been identified in the biogeographic Guiana region. Guyana provides habitats for a variety of fauna and is deemed to have one of the richest mammalian faunas of any comparably sized area in the world. There are nearly 1,200 vertebrates, of which 728 are birds, 198 mammals, 137 reptiles and 105 amphibians. Of these, CITES currently lists 44 as in danger of extinction. The avifauna in its habitat in the widespread rain and seasonal forests is known to be rich. Nevertheless, there is a paucity of data about the country's reptiles, amphibians and fish; only three surveys approved by CITES have been done in the past ten years: two for caimans and one for boid snakes.

Guyana's forests have abundant wildlife, but there are no reliable population surveys of commercial and other species. Therefore, there is no measurement of whether wildlife harvests exceed sustainable levels, nor whether and when closed harvest seasons should be established annually for commercial species. Wildlife exports were estimated at US$800,000 in 1992, mostly from the sale of birds, particularly parrots and macaws. Guyana is the fifth largest exporter of birds in the world. Besides the seventeen authorised exporters of wildlife, the trade is a major source of income for thousands of trappers (mostly Amerindians), intermediaries, carpenters who build holding stations, cages and export boxes, and farmers who provide food for the birds and animals. Operating without the benefit of population or biological surveys, Government devised, with CITES and the World Trade Monitoring Unit, an empirical export licence formula to set export quotas.

In December 1992, Government decided not to issue any licences for wildlife exports in 1993 as a prelude to passage of the Conservation of Wildlife Bill. Despite a general lack of public awareness about wildlife issues, there is a strong lobby in and outside Guyana against the wildlife trade, a trade that generates considerable export revenues as well as domestic income and employment.

Government has put in place a system for training wardens and monitoring the wildlife trade, in order to fully comply with the pertinent international conventions. Arrangements also have been made with the San Diego Wildlife Park for continuing assistance in this area. Given these advances, Government decided to lift the wildlife export ban in September 1995.

The sand and shell beaches between the Moruka river mouth and Waini Point on the northwest coast are the nesting grounds for four species of marine turtles: the leatherback, *Dermochelys coriaceae*; the hawksbill, *Eretmochelys imbricata*; the olive ridley, *Lepidochelys olivacea* and the green turtle, *Chelonia mydas*. Despite legislation to protect the turtles, intensive exploitation of the turtles and their eggs for food has resulted in ever decreasing populations and the threat of their extinction.

To protect the turtles, Government has required that turtle exclusion devices (TEDs) be installed on shrimp nets. In addition, the Government will work with communities of artisanal fisherman to help them develop economic alternatives to hunting turtles and their eggs, and to educate them in the vulnerability of the turtle population to that kind of predation.

Studies on various topics related to biodiversity have been conducted in specific areas by the World Wildlife Fund (WWF), Conservation International, the Smithsonian Institute, Global 2000, and the U.K. Natural Resources Institute. Most of these studies emphasize the need for an established system of national parks, by which the extensive biodiversity resources can be nurtured and protected.

National Parks and Protected Areas

Government is committed to ensuring the integrity of forest systems, the conservation and protection of selected forest areas with high species diversity as genetic reservoirs for the future, the allocation of outstanding natural areas for recreational purposes, and the preservation of the country's historical and cultural heritage. This commitment is confirmed in the NEAP. The National Parks Commission Act of 1977 gives management authority to the National Parks Commission (NPC), but the NPC is under-funded and more oriented to urban recreational parks.

The Protected Areas project in the NFAP was given high priority and pledges of funding at the international round-table in February 1993. Such a system will incorporate ongoing work to identify the places of special natural, scientific, and cultural interest.

The NFAP calls for the protection of the Kaieteur National Park and 14 other natural areas, including a biosphere reserve in the southwest (for which a Guyana Biosphere Reserve Bill has already been drafted) and a World Heritage Site at Mt. Roraima. Conservation International completed a rapid assessment of the Kanuku mountains and the EEC financed a study for the creation of a protected area in the Kanuku Mountains and adjoining savanna areas of the Rupununi region. Consultations with Amerindian communities in these areas will be ongoing as their involvement and agreement will be critical to the success of these protected zones.

Kaieteur National Park is the only legally established protected area in the country. Legislation establishing Kaieteur was passed in 1929 and the Kaieteur National Park Act of 1973 provides the legal framework for its constitution and management.

With no land-use planning in effect, the park is not demarcated on the ground and only one park ranger safeguards the ecological integrity of the entire protected area. As presently constituted, the park comprises the falls, the greater part of the gorge below, and part of the Potaro river above the falls to the south. Currently, the designated area counts with only minimal infrastructure. Full authority over the park is not clearly defined as the GGMC, GFC, and the Land and Surveys Commission each have different responsibilities in the park.

WATERSHEDS

Forest cover in the watersheds helps the infiltration of rainwater into the ground which then charges aquifers; the forest cover protects against flash flooding and soil erosion. Some forest cover has been lost to competing activities, such as bauxite mining, agriculture, and harvesting of fuelwood and poles. The watersheds in the coastal plain and in the sandy rolling lands that supply the conservancies have not been protected or managed for water production.

The forests nearest to urban centres and the coast have been heavily exploited for fuelwood by household and industrial users and have also suffered repeated wildfires from charcoal and agricultural production. The wallaba (eperua) forest of the White Sands peneplain, which has been harvested for telephone and electricity poles, has been degraded to such an extent that regeneration of the original dry evergreen forest is virtually nonexistent. As much as 200,000 hectares are believed to be unable to regenerate spontaneously. Most of the water courses from the watersheds are also subject to competing demands: for drinking and irrigation water and as receptacles for domestic and industrial waste.

Consequently, much of the coastal plain's supply of potable and irrigation water is believed to be polluted. Deforestation may also explain the more frequent and less predictable flooding of the coastal plain. While no clear picture emerges of the extent of the deforestation, potential pressures from logging could have serious consequences for the critical watersheds unless preventive measures are taken ahead of time.

DIVISIONS OF INDIAN BIODIVERSITY

BIODIVERSITY OF HIMALAYAS

Himalayan Flora

Himalayas are one of the hotspots of India. A variety of plant life grows here as a result of varied climate. Both the vegetation as well as wildlife change with altitude and resulting wildlife climatic conditions. Thousands of species of flora and fauna flourish here which adapt to climate predators and other challenges. However many Himalayan species have become extinct and are threatened on the verge of extinction it is directly attributable to the intervention of man in the ecosystem given below is the brief description of tremendous wealth of the Himalayas.

- *Vertical Change in Vegetation*: Himalayan vegetation varies with altitude and climate. As a result variety of trees grow here in the foothill zone, we can find the tropical delicious forests Tropical delicious forests are found in the areas of lower rainfall. They are mainly found in the slopes of the Himalayas trees of these forests shed their leaves once a year. Teak and Sal are common in the areas of delicious vegetation. In the middle altitudes, the temperate forests grow. Pine, spruce, cedar, fir, juniper are common of middle altitudes. Further high, there are coniferous, subs – alpine and alpine forests. Still above these are areas of permanent grasslands and high altitude melons that gradually disappear in the zone of permanent snowline.
- *Horizontal Change in Vegetation*: The Himalayan vegetation is marked by the tropical Rainforests of Eastern Himalayas and alpine forests of central and western Himalayas. The tropical rainforests of eastern Himalayas are dense, evergreen and gloomy which resemble the areas of Amazon basin.

In the cold desert of Trans Himalayas we find only sparse desert vegetation Chir Pine grows throughout the Western Himalayas except Kashmir. Other major types of vegetation in western Himalayas are: Chilgoza or pine nut, maple, ash and oak.

Changes Bought by Nan

Just like many other areas on Earth, the Himalayas have also been encroached by man. With the increase in population, the demands for forest products also grew.

The delicate natural balance got disturbed and forest cover started reducing in size. The man cut a large number of trees to meet his domestic and industrial demands. Forests have also been cleared for agriculture. Many species have become extinct and many others are endangered.

Himalayan Fauna

The animal Kingdom in the Himalayas in as diverse as the plant kingdom. The Himalayan wildlife differs greatly from the wildlife of other parts India. The animals live in different habitats, ranging from deciduous forests to alpine.

In the Himalayas, animals of lower altitude too are adapted to warmer conditions whereas, animals of alpine region can live even in the betting cold in the eastern Himalayas, the climate is warm and moist, so the animals adjust accordingly. Seasonal movements with the change in season many animals shift to lower altitudes in winters in search of food. As soon as the winters are over, they migrate to higher reaches animals of extremely cold areas have thick fur and bushy tails. Animals of higher reaches are adapted to rarefied air, therefore, have larger nasal cavities.

Carnivores of Himalayas are elusive of all the mammals of Himalayas some of them are threatened and are rare like snow leopard. It is due to indiscriminate hunting by man. The lower region is an area of bigger mammals like swamp deer, cheetel, hog deer, barking deer, wild boar, tiger, panther, wild dog, black and sloth bear and elephant. Besides, prominent scavengers like hyena and Jackal are also present.

The common animals in the higher altitude of Western Himalayas are goat, sheep, yak and wild – ass. Ibex is a famous species of goat. Thar and Marknor are also popular species of goat.

In the Eastern Himalayas, wildlife is significantly different from other parts of Himalayas. Some of animals common in the region are red panda, badgers, porcupines, ferrets etc. the popular goats found in the region are serow, goral and takin.

BIODIVERSITY IN TROPICAL RAINFORESTS

The region of tropical rainforests comprises of Eastern Himalayas, Western Ghats, west Bengal and Andaman and Nicobar Islands. All these regions receive heavy rainfall evergreen trees are prominent feature of these forests. In these forests the upper most storey consists of tall trees and makes the canopy that doesn't allow the sunlight to enter. Trees of comparatively lesser height make the second storey.

They remain in shade of trees. This think and dense vegetation provides habitat to a number of animals. These prominent animals are elephant, Nilgiri Langur, Lion tailed macaque, Slender Toris, Malabar civet and spring mouse. Gaint squirrels, civets and bats are also found in the tropical forests. Red panda and golden Langur are typical of the Eastern Himalayas.

Rainforests of Andaman and Nicobar Islands have Characteristic features because they are free from Human interference. Here we can find the most beautiful rainforests of the world more than 200 species of trees grow here some of them are: Padauk, Gurjain, Silver gray etc. the common animals of these islands are: wild Pig, hombill, Andaman teal, Nicobar Pigeon, white beuied sea eagle, Andaman cat Snake, Nicobar legless snake etc.

Mangrove forests in the Sunderban delta in west Bengal are habitat of Bengal tigers. Highest numbers of tigers are found in this part of India other prominent animals spotted are deer, pig, monkey, lizard, water monitor, crocodile, crab and fish. The tigers of these forests can swim in water. It mainly preys on the spotted deer, wild boar and fish Bengal Tigers also attack humans.

BIODIVERSITY IN TROPICAL DECIDUOUS FORESTS

Most parts of India are covered by topical deciduous forests. Trees of this region shed their leaves once in a year. They grow mainly in Peninsular India and Indo-Gangetic Plains. The area under tropical deciduous forests extends from the base of Himalayas to Kanyakumari. The region is home for a variety of flora and fauna.

In the western parts, the region is comparatively dry. The western most part is called the Thar Desert. It receives minimum rainfall. The flora of this region consists of dry tropical, dry mixed deciduous, Thom forests, scrub forests and dry savanna forests. In the extreme west, xerophytes vegetation is dominant. This vegetation has thick outer layer from which evapotranspiration is minimum many of the desert plants are thorny with reduced leaf surface. Cacti and succulents are major plant species of desert area.

Fauna of this region is also diverse the prominent animals are sambhar, wild boar, gaur, Chettal, hog deer, swamp deer or barasingha, nilgai, black buck, elephant, muntjak, common mongoose, wolf, squirrel, hare, wild dog, tiger, leopard, lion, hyena, jackal, jungle cat etc.

Many of the animals are endemic to the region. Animals of desert area are adapted to face the water scarcity and extreme hot conditions. Asiatic wild – ass and black – buck are the main animals of the desert area.

The major national parks of the region are: simplipal in Orissa; ramthambore and Sariska in Rajasthan; Belta in Jharkhand; van Bihar in Madhya Pradesh etc.

AGRO-BIODIVERSITY IN INDIA

India is also centre for crop diversity it is the homeland of 167 cultivated species and 320 wild relatives of crop plants. India's record in agro-biodiversity is again very impressive there are 167 crop species and wild relatives. India is considered to be centre of origin of 30,000 to 50,000 species of rice, pigeon pea, mango, turmeric, ginger, sugarcane, goose – berries etc. and ranks seventh in terms of contribution to world agriculture.

Economic Potential of Biodiversity

Both plants and animals are a source of variety of food products. Wheat and rice are the staple food in India. Farmers also grow vegetables, cash crops, cereals etc. orchards are the source of fruit many plants are a source of medicines.

Several life saving drugs are obtained from them animals like sheep and goat provide wool. Domesticated animals like horse, camel etc. help in transportation activities. Still in many parts of India, fuel wood is the only source of energy. Forests amount to store of huge potential of fuel wood energy.

Biodiversity also helps in making the air clean. Through photo synthesis, plants absorb carbon Dioxide and release oxygen large tracks of forests help in absorbing and decomposing pollutants. They help in maintaining hydrological cycle, protecting and recharging of watershed. It helps in making the water clean.

BIODIVERSITY CONSERVATION

The conservation of biodiversity is of two types: the In Situ conservation and the Ex Situ conservation. Normally, the In Situ conservation is more cost effective however, in some cases the ex-situ conservation has to be adopted.

In Situ Conservation

The major objective of protecting the biodiversity is that the ecosystem and biodiversity are able to flourish and evolve. First of all, in situ conservation requires the identification of areas that are to be protected or that have very high biodiversity it can be done by establishing national parks and national reserves. A reserve is split into patches to provide corridors for the movement of animals.

Various Strategies for in situ conservation are:

- *Surveys:* Data on species diversity populations, locations, extent of habitat, major threats etc. are collected. In our country the 'Ministry of Environment and forests plays a guiding role by pre-paring a list of priority issues and areas.
- *Biosphere Reserve:* A Biosphere reserve is an international conservation designation given by UNESCO. It is created to promote

balanced relationship between humans and biosphere. The non-conservation activities are totally prohibited in the biosphere reserve. As on June 2005, there are 482 biosphere reserves in 102 countries of the world.

- *National Park:* It is a reserve of land that is owned and declared by the national government of a country. A National park is usually located in under developed areas and is protected from human developmental activities and pollution. It is a reserve of land to save exceptional native plants and animals, Scenic beauty of landscape, geological formations etc. At present, there are around 4000 national parks throughout the world. In India, there are 89 national parks that cover 1% of the countries land area.

Ex Situ Conservation

The approach of ex-situ conservation is used to preserve biodiversity in an artificial setting. It includes ways like storage of seeds in banks, breeding of animals in 2005, setting up of aquariums, botanical gardens, research institutes etc.

- *Captive Breeding:* It is the process of breeding endangered animals by capturing them from their natural environment and then breeding them in unfavourable conditions. Later, they are realized back to their natural habitat when threats to the animals is lessened or removed. This activity is being practised in many parts of the world with great success.
- *Botanical Gardens:* The botanical gardens grow a variety of plants purposes. In most cases, they are displayed for public, because they have educative value also many botanical gardens depend on public for funding while some are funded by other institutions or the government. The botanical gardens sent their plant collecting expeditions to various parts of the world and publish their findings.
- *Gene Banks:* Gene Banks is means to preserve the genetic material scientists preserve the plant genetic material in the form of seeds or freezing cuts of plants. The animal genetic material is preserved by freezing the sperms and eggs in zoological freezers. Scientists have thus prevented a gene family from ending.

IMPORTANCE

The increasing use of the term biodiversity is being driven by the fact that, in an ecological context, global biodiversity itself is being lost at an alarming rate. Although it has been shown that the significant global biodiversity loss that has occurred over the timeframe of human existence has not stopped global human population increase, there is clear evidence that biodiversity loss can affect the wellbeing of society and have negative economic impacts.

Biodiversity underpins ecosystem function and the provision of ecosystem services. Biodiversity loss therefore threatens the provision of goods and services provided by ecosystems. Reduction in biodiversity can affect decomposition rates, vegetation biomass production and, in the marine environment, affect fish stocks. It is predicted that a reduction in marine productivity means that fisheries will not be able to meet the demands of a growing global population.

In addition to the gradual decline in environmental function linked to reductions in biodiversity, it has been suggested that there is a risk that at some point a threshold will be crossed and a catastrophe may occur. Research has highlighted that biodiversity loss could rival the problems of carbon dioxide increases as one of the major drivers of ecosystem change in the 21st Century. Whether from environmental collapse or gradual decline in function, our ability to adapt to a changing world may be considerably reduced if the environment on which we rely does not contain sufficient biodiversity to evolve and continue to support our needs.

MULTILATERAL ENVIRONMENTAL AGREEMENTS

In response to the current rate of biodiversity loss, and on the grounds that biodiversity is a common concern for humankind, the Convention on Biological Diversity (CBD) was opened for signature in 1992. As of June 2013 it has been ratified by 193 parties (Governments). The CBD provides a global legal framework for action on biodiversity and is considered a key instrument for sustainable development. Its three main goals are:

1. The conservation of biological diversity;
2. The sustainable use of the components of biological diversity;
3. The fair and equitable sharing of the benefits arising from the use of genetic resources.

The CBD's governing body is the Conference of the Parties (COP). It holds periodic meetings to review progress on the Convention targets, and advance its implementation. To support implementation of the CBD, the United Nations General Assembly declared 2011-2020 the United Nations Decade on Biodiversity and adopted the Strategic Plan for Biodiversity 2011-2020. The Strategy is a ten-year framework for action adopted by signatory countries in 2010 in Nagoya, Japan. It builds on the vision that "by 2050, biodiversity is valued, conserved, restored and wisely used, maintaining ecosystem services, sustaining a healthy planet and delivering benefits essential for all people".

The Strategy calls for all countries and stakeholders to effectively implement the three objectives of the CBD by establishing national and regional targets, feeding into the five strategic goals and 20 global targets (collectively known as the Aichi Targets) outlined by the Strategy. The primary framework for action set forth by the CBD is the ecosystem approach, an integrated strategy for the management of biodiversity resources.

Biodiversity is also at the centre of a number of other Conventions e.g. the Convention on Migratory Species (CMS), the International Treaty on Plant Genetic Resources for Food and Agriculture (Plant Treaty), The Convention on International Trade in Endangered Species of Wild Fauna and Flora (CITES). It is also the subject of a number of associated Protocols such as the Specially Protected Areas Protocol and the Cartagena Protocol.

A new platform, the Intergovernmental Science-Policy Platform on Biodiversity and Ecosystem Services (IPBES), was established by the international community in 2012 and is open to all United Nations member countries.

It is an independent intergovernmental body committed to providing scientifically-sound assessments on the state of the planet's biodiversity in order to support informed decision-making on biodiversity and ecosystem services conservation and use around the world.

MEASURING AND MONITORING BIODIVERSITY

Over the last 30 years, many different definitions of biodiversity have been used. As early as 1992, the year the Convention on Biological Diversity was opened for signature at the Rio Earth Summit, it was noted that the definitions of biodiversity are "as diverse as the biological resource". While the CBDdefinition is commonly accepted, the variety of definitions of biodiversity is particularly relevant when it comes to the scientific measurement of biodiversity. For the purposes of detailed analysis, and the creation of indicators to measure or monitor trends, exactly how biodiversity is defined will influence what is measured.

Biodiversity indicators aim at using quantitative data to measure aspects of biodiversity, ecosystem condition, services, and drivers of change. This advances understanding of how biodiversity is changing over time and space, why it is changing, and what the consequences of the changes are for ecosystems, their services, and human well-being. The huge variety of elements included in the definition of biodiversity results in a varied set of methodologies to measure the natural environment. There is no unified metric for quantitative measurement. The variety of metrics employed include:

- Species richness (number of species);
- Population number (number of genetically distinct populations of a particular species defined by analysis of a specific element of its genetic makeup);
- Genetic diversity (The variation in the amount of genetic information within and among individuals of a population, a species, an assemblage, or a community;
- Species evenness (measurement of how evenly individuals are distributed among species); and

- Phenotypic (organism characteristics) variance, (the measurement of the different between the phenotypes within a sample).

BIODIVERSITY MEASUREMENT AND POLICY

To measure biodiversity is an area of discussion particularly relevant at the science/policy interface. The CBD-mandated Biodiversity Indicators Partnership (BIP) promotes the development of indicators in support of the CBD and related Conventions, national and regional governments and a range of other sectors. Indicators initiated under the partnership are linked to the goals of the Strategic Plan for Biodiversity 2011-2020 and include habitat extent, protected areas and species extinction.

10

Elements and Types of Biodiversity

BIODIVERSITY

The word 'biodiversity' is a combination of two words: biological and diversity. It refers to the variety of life on Earth. Biodiversity encompasses all the living things that exist in a certain area, in the air, on land or in water: plants, animals, microorganisms, fungi. The area considered may be as small as your backyard compost heap — or as big as our whole planet.

Animals and plants don't exist in isolation. All living things are connected to other living things and to their non-living environment (earth forms, rocks and rivers). If one tiny species in an ecosystem becomes extinct, we may not notice, or think it's important. But the biodiversity of that ecosystem will be altered, and all the ecosystems that the species belonged to will be affected.

IMPORTANT OF BIODIVERSITY

The natural environment is the source of all our resources for life. Environmental processes provide a wealth of services to the living world — providing us with air to breathe, water to drink and food to eat, as well as materials to use in our daily lives and natural beauty to enjoy.

Complex ecosystems with a wide variety of plants and animals tend to be more stable. A highly diverse ecosystem is a sign of a healthy system. Since all the living world relies on the natural environment, especially us, it is in our best interests and the interests of future generations to conserve biodiversity and our resources.

Some might argue that some species have become extinct, with no obvious effect on the environment. But the Earth's systems are so complex that we are still learning about environmental processes and resources and the roles they play. The careless loss of any part of the natural environment means that we may never know what use it was or could have been in terms of future technologies, say, or for medical science, or indeed for the health of the planet itself. It's important to understand that environments are constantly changing. A healthy, robust environment evolves and adapts to naturally changing conditions. It is fascinating to observe the far-reaching effects even small changes can make and the importance of genetic diversity for species to adapt, survive and evolve.

Preservation of biodiversity is not necessarily about preserving everything currently in existence. It's more a question of 'walking lightly' on the Earth — a balance of respecting the natural changes that occur and of protecting species and environments from wanton extinction and destruction.

Life on Earth would not be the same if our planet's biodiversity were to be radically affected.

THREE TYPES OF BIODIVERSITY

There are three aspects to biodiversity: species diversity, genetic diversity and ecosystem diversity. All three interact and change over time and from place to place.

Species diversity refers to the variety of different living things.

Genetic diversity refers to the variations between individuals of a species — characteristics passed down from parents to their offspring.

Ecosystem diversity refers to the great variety of environments produced by the interplay of the living (animals and plants) and non-living world (earth forms, soil, rocks and water).

THREATS TO BIODIVERSITY

There are many threats to biodiversity. One of the problems is simply the number of people on Earth.

As the Earth's population grows, and people encroach on more and more land, habitats can become fragmented and degraded. Another threat to biodiversity is the issue of climate change. There is world-wide concern over global warming. Pollution of our air and waterways also affects biodiversity. Overharvesting depletes the Earth's stocks of animal and plant resources and the ill-considered introduction of exotic species also has a negative effect on biodiversity. Many of these threats to biodiversity are the result of human activity. But it is also humans who can make positive changes — adopting attitudes to the Earth's rich resources that will respect and maintain this diversity.

TYPES OF BIODIVERSITY

The term biodiversity includes three different aspects, which are closely related to each other. Following are the types of biodiversity:

GENETIC DIVERSITY

It refers to the variation of genes within the species. This constitutes distinct population of the same species or genetic variation within population or varieties within a species.

Species Diversity

It refers to the variety of species within a region. Such diversity could be measured on the basis of number of species in a region.

Ecosystem Diversity

In an ecosystem, there may exist different landforms, each of which supports different and specific vegetation. Ecosystem diversity is difficult to measure since the boundaries of the communities, which constitute the various sub ecosystems, are elusive.

Ecosystem diversity could best be understood if one studies the communities in various ecological niches within the given ecosystem; each community is associated with definite species complexes. These complexes are related to composition and structure of biodiversity.

Agro-Biodiversity

The agricultural biological diversity more commonly referred to as the agro-biodiversity has been fast emerging as a strong, evolutionary divergent line from the biodiversity, which deals with the life forms at large. It has been specifically recognized to differentiate between concern for ecosystems versus agro-ecosystems, wild forest flora and fauna versus agriculture related plants, reptiles, insects, avian and microbes; in situ conservation of wild forms versus on farm conservation of landgraves and traditional/ primitive cultivars or ex-situ conservation of plant genetic resources, etc.

Agro-biodiversity in a traditional farming system is as follows:

- Rich in plant and animal species
- A wide diversity of niches in the local environment utilized
- Reuse of organic residues, consuming biomass enabled
- Ecosystem functions, such as pest, weed and disease management enhanced
- Locally available resources consumed to an advantage
- Reduction of risk and optimization of resources use
- Associated with farmers time tested local knowledge about resources.

BIODIVERSITY AT GLOBAL LEVELS

It is estimated that there exists 5-30 million species of living forms on our earth and of these, only 1.5 million have been identified and include 3,00,000 species of green plants and fungi, 8,00,000 species of insects, 40,000 species of vertebrates and 3,60,000 species of micro-organisms.

Recently it has been estimated that the number of insects alone may be as high as 10 million, but many believe it to be around 5 million.

The tropical forests are regarded as the riches in biodiversity. According to the opinion of the scientists more than half of the species on the earth live in moist tropical forests, which is only 7% of the total land surface. Insects (80%) and primates (90%) make up most of the species.

Table. Estimated Number of Species Worldwide.

Taxonomic Group	Number of Species
Bacteria	3600
Blue Green Algae	1700
Fungi	46983
Bryophytes	17000
Gymnosperms	750
Angiosperms	250000

STATUS IN INDIA

During the last few years, the subject of conservation of biological diversity has attracted considerable attention at the national and global levels. India is a rich centre of biodiversity and has contributed many economic plants to the world and many useful genes for genetic upgrading of cultivated plants and domesticated animals.

India has a land mass of 329 million hactares with a diversified eco-geographical regions. Almost all types of habitats available in the world are found in India. There are two biogeographical realms in India and it is the confluence of floras and faunas of Africa, Mediterranean, European, Sino-Japanese and Malayan regions. As a result, we have a rich biological diversity.

Plant Species Diversity

At present 1.7 million species have been recorded so far in the world. India's contribution to this record stands at 7 %. Surveys conducted so far have inventorised over 47,000 species of plants and over 89,000 species of animals. Survey and inventorisation of India's biodiversity is still far from complete especially the lower groups of plants and invertibrate animals.

Table. India's Biological Wealth

Plant Taxa	Species	Animal Taxa	Species
Bacteria	850	Protista	2577
Viruses	Unknown	Mollusca	5070
Algae	6500	Arthropoda (Insecta, Crustacea etc.)	68389
Fungi	14500	Other Invertebrates	8329
Lichens	2000	Protochordata	119
Bryophytes	2850	Pisces	2546
Pteridophytes	1100	Amphibian	209
Gymnosperms	64	Reptilian	456
Angiosperms	17500	Aves	1232
		Mammalian	390
Total	**45364**	**Total**	**89317**

Based on the available data, India ranks 10th in the world and 4th in Asia in plant diversity and ranks 11th in the number of Angiosperm species. India ranks 10th in the number of mammalian species and 11th in the number of endemic species of higher vertebrates in the world.

Table. Number of Angiosperm Species in Different Countries.

Country	Angiospermic species
Brazil	55000
Colombia	45000
Ecuador	29000
China	27000
Mexico	25000
Australia	23000
South Africa	21000
Indonesia	20000
Venezuela	20000
Peru	20000
India	17000

Floristic Status

As noted earlier 47,000 species of plants representing about 12 % of the recorded world's flora have already been identified. Comparative statement of recorded number of plant species in India and the world is given in Table.

Table. Comparative Statement of Recorded Number of Plant Species in India.

Plant Taxa	Species		Percentage of India
	India	World	to the World
Bacteria	850	4000	21.25
Viruses	Unknown	4000	-
Algae	6500	40000	16.25
Fungi	14500	72000	20.14
Lichens	2000	17000	11.80
Bryophytes	2850	16000	17.80
Pteridophytes	1100	13000	8.64
Gymnosperms	64	750	8.53
Angiosperms	17500	25000	7.00

Flowering plants accounts nearly 17,500 among 45,000 species of plants. The important economic species includes rice, sugar cane, coix, beans, cowpeas, banana, Citrus, mango, coconut, cardamoms, nutmeg, tea, cotton, jute, colocasia, pepper, ginger, Rhododendron, Jasmines, bamboos, Orchids, betel leaf etc.

Two regions of our country harbours maximum diversity, they are North-East and Sourth-West India. The North- East region is a very active centre of evolution and has diversity for a number of plants like Rhododndron, Camelia, Magnolia, Buddleia, etc. On the basis of distribution pattern of plants, Good, divided plant wealth into 37 floristic zones. With in these zones, pockets of diversity of plants species arose and they were domesticated by human kind during the past 10,000 years. During these years enormous variability was generated because of mutation, recombination and selection process. The result being complex variation pattern in plants. These evolved plants bear little or no resemblance with their ancestors.

Endemic Species

Every major habitat, from areas of heavy rainfall to the dry desert, from coldest to the hottest climatic conditions, from highest elevation down to the sea level is found in the country. India has a rich endemic flora.

Endemism of Indian biodiversity is significant. About 4950 species of flowering plants or 33% of this recorded flora are endemic to the country. These are distributed over 141 genera belonging to 47 families. These are concentrated in the floristically rich areas of North-East India, the Western Ghats, North-West Himalayas and the Andaman & Nicobar Islands.

The Western Ghats and Eastern Himalayas are reported to have 16000 and 3500 endemic species of flowering plants, respectively. These areas constitute two of the 34 hot spots identified in the world.

Cultivated Plants

Table. Active Germplasm Holding and Base Collections at NBPGR.

Crop groups	Active Germplasm	Base Collection Holdings
Cereals	12086	43409
Pulses	38695	22269
Millets & Minor Millets	10349	14488
Oilseeds	19808	14278
Vegetables	12146	5681
Medicinal & Aromatic Plants	870	942
Pseudocereals	4739	736
Tuber Crop/ Spices	2053	-
Forage Crop	4060	-
Horticultural/ Ornamentals	22212	-
Fibre Crops	-	3212
Released crop Varieties	-	904
Reference Samples (Medium Term)	-	53161
Total	**107018**	**159080**

Indian region alone has given to the world nearly 167 economical plants whose centre of origin/ diversity lie in India along with their 320 species of wild relatives and land races. India is considered to be the centre of origin of rice, sugar cane, minor millets, pigeon pea, brassicas, rice-bean, Asiatic vignas, egg plant, banana, citrus, mango, cardamom, jack fruit, jute, edible diascorea, black pepper, seed amaranths, turmeric, ginger, several umbellifers and cucurbits, bitter gourd, colacasia, okra, coconut, bamboo, taro, indigo, sun hemp, gooseberries and many herbal drugs, rhododendron, jasmine, some orchids and betel nut. This remarkable diversity of life-forms in a single country is because of the great diversity of ecosystems. The gene bank of National Bureau of Plant Genetic Resources (NBPGR) has a collection of over 1,59,080 varieties. The details of the active germplasm holding and base collections of NBPGR are given in Table above.

Wild relatives of Crops

There are several hundred species of wild crop relatives distributed all over the country. A major centre for wild rice is the eastern peninsular India, i.e., West Bengal, Orissa and Andhra Pradesh.

The North-Eastern hills and Tamil Nadu hills are rich in wild relatives of millets. Wild relatives of wheat and barley have been located in the western and North-Eastern Himalaya. Table gives the statement of wild relatives of crops recorded so far.

Table. Wild Relatives of Crop.

Crop Wild relatives	No. of
Millets	51
Fruits	104
Spices and condiments	27
Vegetables and pulses	55
Fibre crop	24
Oil seeds, tea, coffee, Tobacco, & sugarcane	12
Medicinal plants	3000

BIODIVERSITY CONSERVATION IN ENVIRONMENT

In the Convention on Biological Diversity signed by many member states at the Earth Summit held in Rio de Janeiro (Brazil) in 1992, explains biodiversity as follows: "Biological diversity" means the variability among living organisms from all sources including terrestrial, marine and other aquatic ecosystems and the ecological complexes of which they are part; this includes diversity within species, between species and of ecosystems.

TYPES OF BIODIVERSITY

Biodiversity is a generic term that can be related to many environments and species, for example, forests, freshwater, marine and temperate environments, the soil, crop plants, domestic animals, wild species and micro-organisms. Basically it can be classified according to three types of diversity:

- Ecosystems and landscapes (habitat diversity)
- Animal, plant, bacterial species (species diversity)
- All genes (genetic diversity)

Of particular importance are the taxonomically isolated species, as they have little similarity to other species and therefore are unique with respect to their genetic structure. These species are often endemic meaning limited to one specific area. Their extinction would mean a greater loss for global biodiversity rather than just the extinction of a species.

Distribution of Biodiversity

In Europe, biodiversity is very unevenly distributed with the least variety of ecosystems and lowest diversity occurring in Northern Europe. Centres of high biodiversity are to be found in the Mediterranean (Italy, Spain, Greece, France) and on the fringes of Europe (Bulgaria, Ukraine, Georgia, Armenia, Turkey), with over 5,000 endemic plant species that occur only in these countries. The Mediterranean is Europe's richest sea in terms of biodiversity.

The Importance of Biodiversity

Biodiversity is often used to draw attention to issues related to the environment. It can be closely related to

- The health of ecosystems.

For example, the loss of just one species can have different effects ranging from the disappearance of the species to complete collapse of the ecosystem itself. This is due to every species having a certain role within an ecosystem and being interlinked with other species.

- The health of mankind.

Experiencing nature is of great importance to humans and teaches us different values. It is good to take a walk in the forest, to smell flowers and breath fresh air. More specifically, natural food and medicine can be linked to biodiversity.

Biodiversity in the Black Sea

The Black Sea region used to be one of the most important areas for fisheries and for food and income for local people. Sturgeons (Acipenser sp.), mullets (Mugil sp.), and mackerel (Scomber sp.) as well as other species have been extensively exploited in this area. However, human activities related to agriculture, shipping and tourism now exert strong pressure on the environment, especially in the northern part of the Black Sea, and are taking their toll on biodiversity and damaging fish stocks.

Threats to Biodiversity

Large amounts of fertilizers were carried to the sea from agricultural areas along the large rivers, such as Don, Dnepr, and Dnjestr. Oil leaked into the sea from ports on the eastern shores and directly from tankers crossing the Mediterranean. Wastewater from cities and heavily visited coastal tourist towns was discharged without any purification.

Consequences for the Ecosystem

All of this leads to pollution and eutrophication meaning the enhanced blooming of planktonic algae due to increased nutrient loads. Planktonic algal blooms lower the water transparency letting only a little amount of light through the water where makrophytes usually grow.

This is how the natural belt of bottom vegetation along the Black Sea coast has been destroyed. For example, the vertical distribution range of Cystoseira spp. decreased from 0-10m to 0-2,5m. The consequences were drastic because the habitat is of major importance as a nursery for spawn and hatchlings of many marine species. This led to a drop in reproduction rates and hence fish stocks.

Biodiversity of the Coastal Zone

Coastal areas are exceptionally productive environments, rich in natural resources, biological diversity and with a high potential for commercial activity. The importance of biodiversity in the coastal zone can be demonstrated by 8 out of the 40 EU listed priority habitats of wild fauna and flora falling into the coastal habitat.

Approximately a third of the EU's wetlands are located on the coast as well as more than 30% of the Special Protected Areas designated under the Directive for the conservation of wild birds. The reproduction and nursery grounds of most fish and shellfish species of economic value also comes from this area, which accounts for almost half of the jobs in the fisheries sector.

Pressure on Coastal Biodiversity

Coastal areas are increasingly vulnerable to stresses from both human activities and the forces of nature. The complexity of human activities, natural systems and ownership in the coastal zone, requires an integrated management scheme to allocate coastal resources efficiently and minimize environmental degradation. Choices have to be made between competing uses and limits of resource exploitation if escalating conflicts and resource degradation are to be avoided.

The attitudes of community and industry to the use of biological resources should change from the 'maximum yield' approach to one of 'ecologically sustainable', which recognizes the need for conservation of biological diversity and maintenance of ecological integrity.

Integration of management regimes within and between different sectors to meet environmental, economic and social objectives must be realized in order to achieve sustainable development.

Integrated Approach

Integrated policies will also provide the opportunity for all the people to accept responsibility for their actions and the impact they may have on biological diversity. The development of integrated policies for managing the coastal resources is necessary:

- To coordinate activities within and between all levels of government;
- To ensure that full social and environmental consequences (and costs) of development activities are considered;
- To ensure that the public interest is properly taken into account.

ASSIGNING ECONOMIC VALUE TO BIODIVERSITY

Biodiversity refers to the diversity of life in all its forms and all its level of organisation, not just plant, animal and microorganism species. At its most elemental level, biodiversity, encompasses the varied assemblages of organic molecules that comprise the genetic basis of life. On the other end of the spectrum there are biomes-the vast stretches of tundra, desert, forest, ocean, etc. In between come-population, race, sub-species, community, ecosystem.

Why is Biodiversity Important?

For many people the very questioning of the worth of biodiversity is illicit. They would argue that humankind has a moral obligation to conserve biodiversity, an obligation that comes with the fact that humans have the capability to destroy much of that biodiversity. Others find a religious support for such a view there is some stewardship responsibility on behalf of some deity. One problem with these moral views is that they often conflict with other moral views about, say, the right to earn a living, the right to have access to basic needs such as food, cloth and shelter, and so on. If conserving biodiversity conflicts with those rights, then some 'meta-ethical' principle is required for deciding which moral view should prevail. One aspect of the process of changing popular perceptions about biodiversity resources is to show that the sustainable use of biodiversity has positive economic value and that this economic value will often be higher than the alternative resource uses which threatens biodiversity.

Valuing Biological Diversity Conservation

The issue at hand concerns the measurement of natural capital, great difficulties arise in assigning values to natural capital precisely because the market in which it is traded are limited. In the past few decades a number of techniques have been developed to place monetary value to the benefits of conserving an area for biodiversity. Some rely on market prices of related goods and services both to value benefits and to estimate costs, while other rely on survey based approaches to infer values.

Hedonic Pricing Method

Property values are affected by a number of variables including environmental quality. After excluding other variables, including environmental quality, the residual price difference can then be ascribed, at least theoretically, to differences in environmental quality, *e.g.*, the increased value of property located next to natural areas. According to a report by Nelson (1982) traffic noise, measured in Leq (equivalent continuous sound level), a one unit change produces property price depreciation of 0.5-1.0%.

Ravel-cost Method

This approach looks at the pattern of recreation use of a natural parks and uses this information to derive a demand curve to estimate the total amount of consumer's surplus. For example, travel behaviour reveals that Costa Rican visitors are willing to pay USD 35 per household to visit a tropical rainforest site in Costa Rica. Foreign visitation is likely to be worth far more than domestic, as foreign visitors have higher travel costs.

Contingent Valuation Method

The Contingent Valuation Method (CVM) bypasses the need to refer to market prices by asking individuals explicitly to place values upon environmental assets. This is also referred to as an expressed preference method. An interesting advantage of the CVM approach is that it can, in theory, be used to evaluate resources, that people have never visited personally *e.g.* the Antarctica which people are " Willing to Pay" to preserve but would not in general ever want to visit.

Consumptive Benefit Method

There are a number of products which can be harvested on sustainable basis and marketed. For example, a probable number of higher plant species, which are widely used as basis for pharmaceutical drugs, is some 500,000. According to Pearce and Moran (1994) economic value of these plants can be approached by looking at:

(i) The actual market value of the plants when traded,

(ii) The market value of the drugs of which they are the source materials,

(iii) The value of the drugs in terms of their life-saving properties, and using a value of a 'statistical life',

(iv) The lost pharmaceutical value from disappearing species.

As there is absence of 'global markets' in the benefits of biodiversity, the developing countries face major problems of appropriating the global benefits of sustainable use of biodiversity. As long as these global values cannot be captured by host countries, biodiversity will be a risky investment in many contexts.

SUSTAINING AGRO-BIODIVERSITY

Nations were required to evolve a *sui generis* system for varietal protection, go in for patenting or adopt a combination of both, but not in isolation from the broader goals set out in agenda 21 of the CBD. Such a *sui generis* system, in absolute sense, provides an institutional mechanism and legislative protection to manage and enrich *inter alia* the agro-biodiversity, within the national territory. It would give due opportunities to on-farm conservation for sustainable use besides protecting and safeguarding the interests of those who conserve and maintain the vital components. Issues that the nations must consider, as per the requirements of CBD, include to plan and implement in an effective legal and/or regulatory frame work which would ensure no harmful effects even on territory beyond national borders upon consequent use of various components of agro-biodiversity; the regulation of exchange of germplasm is done in conformity with the charter of the UN and the international law, and that an intellectual property protection is provided to plants or plant parts in other member countries, etc.

Nations may be required to maintain two contrasting agro-ecosystems. First, the traditional agro-ecosystems encompassing and protecting biological diversity therein along with its biotic, abiotic and environmental components providing/maintaining due opportunities to sustainability and the micro-evolutionary processes.

The cost of maintaining such traditional agro-ecosystems, through imparting awareness and providing suitable compensations to the local folk who would practise it, is bound to be high but looking into the indirect impacts of such conservative mechanisms particularly on the climatic equilibrium, the CBD appealed to Global Environmental Fund (GEF), in one of the decisions made in the third meeting of the Conference of Parties (COP-III), to make liberal financial grants for the above purposes pending settlement of the issue of establishing a Global Biodiversity Fund, which is being strongly advocated by several partners and fora. Lessons could also be learnt from the more recent indigenous knowledge and social framework of traditional monospecific cropping systems so as to help maintain the ecological equilibrium in advance, highly productive, intensive agro-ecosystems. Funding provisions may specifically be made for conducting research on these aspects.

Sustaining Alternate Ecosystems

A concern has been shown that the traditional agro-ecosystems should be conserved along with diverse evolutionary forms of cultivated plant species as well as farmers' conscious selections, particularly in the areas of origin and diversity of crop plant species and their relatives. This, obviously calls for concerted efforts towards developing a better understanding in terms of the following:

i. classification, indexing and inventorisation of traditional agricultural systems, still in vogue, with respect to their ecosystems, crop combinations, seasons and/or systems of cultivation;
ii. potential core area(s) under each, respective system or sub-system; and
iii. the custodian community.

At the same time, guiding and maintaining diversity in the advance agro-ecosystems through research, extension and legislation is considered essential to tackle several problem situations, such as, biotic virulences, abiotic stresses and adverse climates. Rapid appraisal systems are, therefore, essential so as to maintain and guide requisite diversity in these agricultural systems to enable fast detection and appropriate replacement of the varietal components, such as those succumbed to diseases/pests/abiotic stresses or those which have been surpassed by more productive types in research. The creation and maintenance of agro-biodiversity could be ensured through multi-crop, multi-varietal mosaic cropping which shall mimic the co-existence of cultivated plant species and the within-species diversity in the traditional agro-ecosystems. Also, the maintenance of the latter through awareness generation and providing suitable compensation to the local people who practise it, would be essential for an effective management of agro-biodiversity.

In Situ on-farm Conservation

There is a demanding need to enumerate agreed scientific principles and practices for 'on-farm' genetic resource conservation. It includes usable classification and nomenclature systems for landraces, as well as methodology for their characterisation. Due to lack of available information of this type, it is imperative that studies should be carried out on these aspects side-by-side while implementing 'on-farm' conservation programmes. A better understanding of the temporal and spatial movement of genes and varieties into and out of cropping systems and the causes of loss or recession of some varieties or traits over others should be highly useful in further setting the pace and direction of crop evolution. Adequate studies of some of the traditional practices and customs have not been carried out in most of the agro-biodiversity rich areas. Systematic surveys and collection of data on the traditional knowledge base, including methods to maintain healthy populations and seed, require an inquisitive follow-up and indexing.

The 'on-farm' conservation is generally associated with traditional agro-ecosystems and small scale farming although the farmers' participatory approach has been more advocated to help select suitable high yielding varieties (HYVs) for the advance cultivation systems. The large-scale commercial farms also have options to promote agro-biological diversity through crop rotations, intercropping, cover crops, green manures, paddy-cum-fish culture, apiculture, etc. The seed requirements for such situations would depend upon the adaptability, preference and acceptability of component HYVs for each sub-situation. Such practices need to be standardised as per diverse agro-ecological zones, so that sustainability of seed component is ensured under a dynamic intensive cultivation system. The more likely negative impact on sustainability of seed under the intensive cropping system may occur due to successively superior varieties being evolved at frequent intervals, the imminent displacement of modified agro-ecosystems which were created by cultivating such superior HYVs, the evolution of virulences or other environmental factors, such as, off-site pollution or accumulation of certain harmful gases like methane or ozone, etc. which need to be addressed by well planned, self-generating mechanisms.

Domestication of individual crop plant species is unique to agro-ecological sub-regions and localities and corresponds to cultural and ethnic diversity, including food habits, habitats, climates and available crop plant species. There is a purposeful need to maintain their representative seed 'on farm' because re-building new varieties for similar situations with same amount of adaptability and phenotypic plasticity would require strenuous planning, breeding efforts and resources, including funds and time. Capacity building for 'on-farm' conservation of traditional 'types' requires formulation of suitable policies, sectoral and cross-sectoral plans and programmes for implementation and also interim financial mechanisms. Recognition of the associated local knowledge, innovations and informal practices would be needed to promote traditional farming systems, whereas establishing community seed banks in cooperative or private sectors, developing and maintaining cultivar/landrace registration system, etc., is required for smooth replenishment. The recognition of 'Farmers' Rights' is important along with formulation of implementable laws and regulations so as to ensure availability of the much needed 'seed' with due compensation or assurance of benefit sharing to its holders. Country level implementation of Registration of Plant Germplasm has been implemented in India by the ICAR and the planning/action related to other aspects is also initiated. Compiling case studies for conservative, low external output based production systems through incentives for developing environmentally sound technologies for conservation, replenishment and sustainable use of biological diversity are issues which lack direction and, therefore, need sincere input from all concerned under the *sui generis* situations.

Geo-sectoral Approach

Contiguity of adjoining areas across the geo-political barriers, representing similar genetic diversity for particular species genepool, can be attributed as a basis to conserve, utilise and share the benefits of the *in situ*-on-farm conservation. Safeguarding the regional 'gene' interests could be ensured through sub-regional consortia and regional networking. The increasing role of Global Forum and the regional fora in strengthening relationships among the national agricultural research systems (NARS) and between various NARS and the Commonwealth Group of International Agricultural Research Institutions (CGIAR), could be viewed as new initiatives in desired direction which need to be pursued. Technological and technical needs can be gradually fulfilled for innovating, standardising and researching into the rapid appraisal systems for agro-biodiversity management using and strengthening the information network(s), as well as database system(s).

Due to the recent global interest for germplasm exchange and concern for protection of Plant Breeders'/Farmers' Rights there is urgency that the countries develop a regional consensus by way of the following:

i. By entering into agreement for undertaking joint explorations in the region for collecting genepools of crops for which the region is centre of origin;
ii. By adhering to effective plant quarantine so as to maintain an effective mechanism for exchange of risk-free germplasm;
iii. By developing *sui generis* methods of protecting the breeders' and farmers' rights by each country considering its own unique situation such that the benefits accruing thereof are also available to the other nations of the region; and
iv. By instituting funding mechanism and ensuring financial support for the regional PGR activities. It is pertinent to emphasise that the seed industry be facilitated and the security of certain vital components ensured.

Consequent upon the adoption of CBD and also implementation/ enforcement of General Agreement on Tariffs and Trade (GATT)/World Trade Organisation (WTO) provisions, new issues and challenges of global significance have opened up. Accordingly, the sovereign rights of the nations on their bio-resources have been specifically recognised. As a consequence, it has become responsibility of the nations to conserve these resources and ensure their availability for their systematic utilisation on sustainable basis. Coupled with this, an equitable sharing of benefits accruing from the use of plant genetic resources (PGR) is an important issue that influences the other vital issue of access to PGR. The funding support from GEF has yet not set its pace which emphasises the need to set up a Global Fund which could *inter alia* help shape and refine the action points on seed security and the seed industry.

There is an urgency to address the identified areas in harmony with the other institutional mechanisms in the fray. Recognising that the gene rich countries have a potential to release naturally evolved variability suited to several diverse situations, there is an adequate need to match with a generous global commitment towards the funding requirement to sustain the traditional/ naturalised diversity on-farm at certain representative sites located within the diversity-rich areas of the world. Some activities might include, but not limited to:

i. Regional exploration teams comprising experts from countries involved could be assigned survey, collection and bio-prospecting of agro-biodiversity. Whereas the hitherto unexplored bio-resources should be collected, evaluated, documented and shared for immediate use besides long-term conservation in the gene bank, the well adapted and popular cultures should be identified along with their companion crops/varieties and the cropping systems, for their on-farm conservation;
ii. Seed multiplication, testing and distribution programmes should be launched for the traditional varieties/folk cultures/landraces which have edge over other co-cultivated variable forms and the farmers provided with critical inputs, either as subsidised or free-of-cost;
iii. Farmers should be encouraged to plant 'rest of the traditional cultures' in a small block of their cultivation field by ensuring seed of all the diverse component cultures from the agency/community seed bank. The produce should be re-purchased at the premium rate of seed rather than commodity;
iv. In order to develop regional capabilities/capacities, the human resource orientation and development needs to be given top priority, giving equal weight to educating both genders;
v. Existing gene bank facilities should be utilised for long term conservation of seed representative of diverse types for posterity. The satellite, medium term facilities, on the other side, should be used to maintain active forms with a view to replenish the lost or diminished types in a catastrophe/crisis; and
vi. Exchange of germplasm seed needs to be regulated by strict phytosanitary regulations and quarantine measures so as to minimise risk from new pathogens/pests prevailing outside the region. Joint ventures in bioprospecting, biosafety and biotechnology, including cryopreservation, *in vitro* conservation or handling of transgenics/ genetically modified organisms (GMOs) need to dwell upon principles of sustainability and synergies/mutual cooperation. All such activities require adequate funding support which should be catalysed and facilitated as per need of the hour.

The damage being done to land, water, flora, fauna and the atmosphere altogether poses the challenge of future food security. It is the human interference that created the crop plants from the wild to their present day commercial forms. Their agro-biological features and dispersal mechanisms have changed drastically since then. The commercial forms are absolutely dependent on human beings for their perpetuation and survival. It is, therefore, appropriate to look into the interests of the seed producers of the required/ desired types. The management approach would require monitoring of the varietal component at frequent intervals so that the balance could be maintained between the interests of the seed producers and the phyto-ecological requirements of the varietal components. Involvement of both governmental and non-governmental bodies and the individuals, as also the genetic resource and patent literacy, constitute an essential elements of systematic sustainability and developmental plans.

FORESTRY AND BIODIVERSITY

The National Development Strategy provides relevant information on the forestry sector in Guyana. Of the total estimated forest area of 65,000 square miles (169,000 km^2), 34,000 square miles (87,800 km^2) lie within the gazetted State Forest boundaries.

The forest resources are characterised by the following:

- Heterogeneity - there are over 100 tree species
- Relatively few species that are of commercial importance
- The small average size of forest trees
- Most of the commercial timber species are dense heavy hardwoods, which according to the Forestry Sector National Development Strategy, make up an average of 80 per cent of the exploited species.

Approximately 2.5 million hectares of forest lands are considered not exploitable for timber with present technology. The total volume of standing timber in the exploitable forest area is estimated at more than 350 million cubic metres, of which hardwoods account for just less than 90 per cent. An area of 3.6 million hectares of forested land is accessible for exploitation, and approximately 2.4 million hectares have been allocated for harvesting. Many rivers that could provide access to the forests are not navigable because of waterfalls and rapids. The forests that are closest to navigable water have been logged for commercial species.

The forestry sector employs about 20,000 people and produces some 218,000 cubic metres of timber annually. Foreign exchange generated by forest products exports doubled during 1994 to approximately US$15 million. The sector is again set to double its export earnings by mid-1996 if plywood exports reach and maintain their target of 10,000 cubic metres per month.

There are seventeen (17) concessionaires operating under Timber Sales Agreements (TSAs). The TSAs are issued for areas >60,000 acres (24,000 ha). The large concessionaires have significant investments in plant and equipment and convey exclusive harvesting rights for 15-25 years. Most of the concessionaires concentrate on logging the endemic greenheart (*Ocotea rodiaei*), and eight of the concessionaires export regularly. All the concessionaires have integrated logging and sawmill operations in the interior, concentrated in the Essequibo and Demerara areas. Small scale loggers have operated with State Forest Permissions (SFPs), generally one year in term, which grant rights to a specified volume of timber within a prescribed area. There are at present 486 SFPs for areas <20,000 acres (8,100 ha) covering approximately 4,126,230 acres

(1,669,801 ha). SFPs were usually issued to smaller operators who produce for the domestic market. About 1.2 million hectares have been covered by SFPs and most of the log production flows to small sawmills concentrated in the Berbice and Demerara areas. Minor forest operators also worked with SFPs in the dry evergreen forests of the Demerara area close to Georgetown, Linden, and the populated coastal settlements.

Monitoring of forestry operations by the Guyana Forestry Commission (GFC) is only nominal because of shortages of staff and equipment, limited funding and absence of a good data base. Large scale concessionaires, small scale loggers, and minor forest products operators still pursue forestry activities with virtually no monitoring by the GFC. The environmental impacts of logging vary with the size of the concession and the conditions attached to the logging permits. Each stage of the logging activity has a separate impact on land, soil, and water. Because of lack of monitoring and regulation, the environmental impacts of logging activities can only be described in qualitative terms, and the sources often disagree as to the seriousness of the various impacts. The following are perceived to be the main impacts from logging activities.

Exploitation of the Greenheart

Greenheart is the best known timber species and is of such intense commercial interest that foresters do not harvest other species in those areas where the greenheart is most concentrated and of best quality. It accounts for only 1.5 per cent of the stand merchantable volume but provides nearly 40 per cent of the total volume of roundwood production. Because of lack of monitoring, conclusive evidence of over-harvesting of the greenheart is not available, but ongoing studies may reveal the true status of this species.

Until recently, efforts to commercialise other species in export markets, to take the pressure off the potentially dwindling greenheart resource, have been only partially effective. Only smaller producers selling on the domestic market have harvested a significant proportion of other species. The high cost of shipping from Guyana constrains exports of species other than the greenheart to North America. Such costs are higher than shipping these same products from Africa and the Pacific. In the case of the greenheart, strong market demand overcomes these factors. Port facilities are also inadequate (though they are being improved), causing costly delays in loading and only small vessels can take a full load. Recently, certain private sector initiatives suggest that marketing possibilities for other species and for non-tree forest products may have improved. The largest and most recent concessionaire, the Barama Company Limited, has established an integrated logging and plywood production system based on harvesting about twelve plywood species. Barama's operation represents a technological and commercial breakthrough for the utilisation of large quantities of lesser known species and more intensive utilisation of forests. The company contracted the Edinburgh Centre for Tropical Forests (ECTF) to undertake monitoring of its environmental impacts and research to benefit its operations.

Impacts on Forest Ecosystems

Concessionaires are required to submit Forest Management Plans (FMPs) as a condition of the TSA, but they have not observed this requirement in the past. Therefore, the long term sustainability of the selective harvesting system used is unknown. Several actions of loggers provide cause for concern. Careless felling and extraction due to under-skilled, poorly supervised, chain saw operators and skidder drivers can result in unnecessary damage to the remaining stand, which is likely to result in lower harvests next time around. Seed trees are not usually retained, wildlife important to seed dispersal are not safeguarded, protective buffer strips along watercourses are not always respected and production demands defer to the speed of the operation rather than to selectivity of harvesting. With one exception, there are no restricted areas within logging concessions that conserve representative areas of productive forest types in their unlogged condition. Consequently, comparison of the ecological condition of the logged forests with the unlogged, once the first full cutting cycles are completed, will not be possible, thereby hampering monitoring of compliance with the FMP.

Impacts on Soils and Rivers

The main impacts of forestry exploitation on rivers and streams are: increased turbidity caused by soil erosion from logging, increased BOD from the discharge of organic waste from sawmilling, and oil pollution from the discharge of petroleum products.

GFC has already issued regulations to protect rivers from the effects of logging wastes, which are automatically included in each FMP. The regulations require the maintenance of 50-metre buffer strips of forest along the river banks. Nevertheless, in some cases when such buffer strips are established, the river banks are demolished by the missile dredges of largely unregulated gold miners. Consequently, the logging waste often slips into the river causing turbidity, with consequent adverse impacts on aquatic life and on the navigability of the rivers. GFC has not yet recommended procedures for the safe storage of oil and fuel away from watercourses nor does it have the capacity to monitor such storage effectively. Proper disposal of sawmill waste has long been an intractable problem. About 60 per cent of the average saw log remains in the yards as waste in the form of slabs, ends, and sawdust. Most mills heap this waste on the riverbanks and it eventually spills over or is pushed into the water, raising BOD levels and hampering aquatic life. The Essequibo area is estimated to generate 50-60 thousand cubic metres of sawmill waste each year. Recently, some concessionaires have started exploring the possibility of utilising the waste as a source of energy. (A pilot project (No. 34) for such conversion was proposed in the NFAP.) Currently, two steam producing boilers fueled by sawmill waste power the Willems Timber and Trading Limited mill at Kaow Island. The Company has also experimented with producing charcoal from sawmill ends in portable kilns. A. Mazaharally Sawmill Limited has acquired a turbine generator to produce electricity from its sawmill wastes. Current efforts at using sawmill waste as a fuel have made little impact on the problem of river contamination, but there may be scope for the conversion of sawmill waste into briquettes using wood densification technology. Forestry exploitation, as practised in here, does not lay bare large expanses of soil or lead to serious soil erosion and land slides at all the logging sites. For economic reasons, concessionaires avoid logging in excessively wet conditions, when the soil is most prone to erosion. The forestry research programme at the Barama concession includes studying the impact of forest operations on soil properties, with a specific focus on soil compaction, erosion and loss of fertility. Tropenbos has been conducting an ecological and forestry research study at the DTL concession at Mabura Hill on the impacts of logging activities on soil properties. The results of such studies are not yet known but will be made available to GFC so that the findings can be incorporated into forestry policy.

Forest Fires

Incidences of illegally set forest fires have caused losses of forest resources. The principal causes of such fires are untended charcoal burning pits and vandalism. The damage from these forest fires could be contained if specific areas were designated for charcoal burning at each camp site and if permittees of the SFP, TSA and WCL were required to report, and, where possible, to attempt to put out fires within forested areas. Monitoring, especially during the dry season, could be the dual responsibility of concessionaires and the local communities.

Forest Management

The management of the forests has not been sufficiently studied to generate a proven silviculture system. In-depth inventories are lacking as are data on growth either in the undisturbed natural forest or in logged-over areas.

The Government emphasised its interest in promoting sustainable forestry development in Guyana through the preparation of the National Forestry Action Plan (NFAP) in 1989. A round-table meeting of the donor community in February 1992, resulted in donor pledges for ten priority projects from the NFAP totalling US$8.6 million. Implementation of the NFAP has continued with the recent appointment of a national coordinator. CIDA agreed to fund an Interim Forestry Project (IFP) that began in 1989, as a bridge between NFAP preparation and implementation. Meanwhile, with assistance from Germany, Government is proceeding to formulate and implement a forest land-use policy, related legislation and an appropriate administrative structure to ensure the sustainable use of its forest resources. As a further expression of its commitment to protection of the tropical rainforest, the Guyana Government advised the Commonwealth Heads of Government Meeting in October 1989, that the country would set aside 360,000 hectares of its Amazon rainforest for a pilot research project under Commonwealth auspices. Since then the Commonwealth Secretariat and the Global Environment Facility (GEF) have begun to finance the establishment of an International Centre for Research and Training for the Sustainable Management of Tropical Forests in the Iwokrama rainforest. This research project will investigate ways that the tropical rainforest can maintain desired levels of biological diversity while supporting economic activity.

FOREST MANAGEMENT

Although silviculture and management may be looked upon as two branches of forestry, the one mainly technical and the other economic, the two are interdependent and must always be coordinated. While silviculture may establish the ideal possibility of production within a forest, management determines the degree to which this can be realized through regulation and within the bounds of good business practice. Forest management may be defined as the application of business methods and technical forestry principles to the operation of a forest property. Compared with other crops, a forest requires a relatively long time between its establishment and its harvesting. Revenue from an unmanaged forest may accrue only at fairly long intervals. Managed forests are organized to ensure a sustained yield of the forest crop in perpetuity. The income should include a reasonable profit on the investment, annually if possible.

Sustained yield of timber depends upon the systematic reproduction of a crop as it is harvested. It also requires that the timber be cut annually or periodically in such quantities that a continuous and fairly uniform yield will be provided throughout the rotation (the period required to grow a crop from seed to maturity). The principle of sustained-yield management may be illustrated with reference to a theoretical forest, usually termed the normal forest, in which the stands are stocked with the maximum amount of wood that the site will produce and from which the same amount of wood product may be harvested annually forever-namely, the amount that grows annually on the whole acreage.

Consider, for example, an even-aged forest that is to be harvested when fifty years of age. To have fifty-year-old timber every year, there must be a series of fifty stands of, respectively, one, two, three and up to fifty years of age. Each year one stand becomes mature for harvesting. This stand in volume represents its own increment or growth for fifty years; also it is equal to the growth of one year on the fifty stands making up the whole forest. Accordingly, each cut is equal in amount to the current year's growth. It is assumed that each stand is reproduced as cut and that all the immature stands receive the silvicultural treatment necessary for most satisfactory growth.

An alternative method, for an uneven-aged forest, is to cut over the entire area periodically, removing at each cut only the volume corresponding to the growth during the period between successive cuts. In either case, one has permanent sustained yield from a forest characterized by normal distribution of age classes, normal increment and normal growing stock. There is, of course, no such forest. Every forest is abnormal, having either excess or deficiency of area in the older age classes and the opposite in younger ones and in general a deficiency, to a greater or less extent, in volume per acre in the stands of different ages. The abnormalities in age-class distribution can be removed only by adjustments in the areas cut over in successive years, while deficiencies in volume, caused mainly by inadequate reproduction, can be made up through improvement in silvicultural practices. The normal forest is an ideal, a standard toward which forests should gradually be developed. Such intensive management requires accurate information about tree species and sizes, total and merchantable volume, age-class distribution, growth conditions, site classes and rotation ages, as well as about such factors as topography and drainage, which affect the efficiency of operations. Such information is obtained through enumeration or inventory surveys, supplemented and improved in recent years by aerial photographs. On the basis of this information, a management plan or working plan is evolved, which sets down the objectives of management and how they are to be attained and allocates the kind and amount of timber to be cut each year or period of years, at least for the first part of the rotation.

LOGGING

Logging engineering is concerned essentially with the felling and removal of timber from a forest area. Even where the sole purpose is liquidation of the resource, logging techniques may be developed to a high degree of efficiency, as they have a direct and important bearing on the cost of extraction. In forests under sustained-yield management, logging methods must be developed to meet the requirements of the management plan as well as to maintain operational efficiency. Logging everywhere presents many technical problems, but particularly where the forests are difficult of access, distances are great and the topography is rugged. In many forest regions logs are transported long distances and at relatively low cost by natural water systems and in the northern forests winter snows often facilitate the hauling of forest products. From the point of view of forest conservation, perhaps the most important logging objectives are:

- To prevent damage to the advance growth and the residual stand,
- To help prepare the site for future tree crops (e.g., by soil scarification),
- To ensure minimum waste of wood in the trees cut.

In recent years mechanical equipment has been increasingly employed in both the felling of trees and the transportation of logs.

UTILIZATION

While it is not usual for those responsible for the growing and harvesting of timber crops to be directly concerned with their conversion and manufacture into usable products, there should be a close liaison between the two operations of forest production and wood utilization. Wood has inherent qualities that make it particularly suitable for many purposes—good strength in relation to weight, pleasing appearance, insulation against heat and sound, ease of fastening with nails, dowels, screws, or glue and case of working with relatively simple tools.

However, wood is by no means a homogeneous material: it differs from one species to another in structure and physical properties and hence in the use for which it is best suited. Within environmental limitations it is desirable to grow the species that will best meet the requirements of local industry and available export markets. Furthermore, to make the fullest use of the timber crop at all stages of development and to minimize logging and manufacturing waste, it is important to co-ordinate as closely as possible forest operations and the wood-using industries and also to bring about a closer integration of the wood-using industries themselves — as, for example, the sawmill and pulp industries.

Forest Policy

In many countries forests are of vital importance to the over-all economy, not only in relation to industrial development, but also in providing benefits which may be difficult to evaluate in monetary terms. For this reason and also because of the long-term nature of the forest enterprise, most countries seek to ensure the perpetuation of their forests and to safeguard the interests of present and future generations by placing the forests under some degree of government control. In some countries the government retains ownership of a large proportion of the forested land and leases it to the wood-using industries for timber extraction: provision may be made for some degree of forest management by regulation or through agreement. Other countries have granted or sold much of their forest land to private interests, though considerable areas have sometimes been retained by the state to provide for timber reserves, recreational facilities and the protection of drainage areas; in some cases there is considerable public control of forest management on the freehold lands.

Elsewhere long-term forest management is entirely neglected and forest policies, if any exist, may be concerned only with the method and intensity of exploitation. Progress in the development and practice of forestry varies widely throughout the world. While forestry is well advanced in much of Europe and in certain tropical regions, only a small proportion of the accessible forests of the world is under intensive management today.

In some regions man still depends on the forests for his food and shelter and contributes little toward their development; in many countries man long ago destroyed the forests and has done little to restore them; elsewhere uncontrolled forest exploitation is still in evidence and proper protection and management lie in the future. In North America we are still in an early stage of forest conservation. In recent years, however, we have begun to see the forests as an integral part of our economic structure and to take steps against the destructive exploitation that has threatened to exhaust them.

FOREST ECOLOGY AND CONFLICTS

The crisis in tropical forest resources is today recognised as the world's most severe ecological and economic crisis. This is primarily because of the linkages between the genetic resources and diversity of tropical forests and food security of the entire world, and the linkages between the ecological stability of tropical forests and the economic well-being of the majority of the world's people who live in the tropics, in what is called the Third World.

These Third World countries, the erstwhile colonies of the industrialized coun tries, provided natural resources on which the industrialization of the latter was based. Both industrialisation and economic growth in the colonial and post-colonial periods have been based on the reckless exploitation of tropical forests. Today, the cumulative impact of this over-exploitation has led to critical and almost irreversible ecological degradation.

The famine in Africa and other arid regions has shifted the world's attention to the high ecological and social costs of tropical deforestation. It has become the central concern of governments, development agencies and ecology movements. Yet the focus on tropical deforestation and its reversal is not automatically translated into the protection of tropical forests and those who depend on them for survival. As long as misconceptions about the nature of tropical forest ecosystems prevail, as long as conflicting demands are ignored and as long as the causes of tropical deforestation are inaccurately located, this degradation cannot be arrested. As a result, tropical ecosystems will continue to be degraded, the survival of people of the Third World will continue to be threatened, and forest conflicts will continue to grow. Normally, forests are identified only with the economy associated with the commercial-industrial use of forests. lowever, the crisis of tropical forestry needs to be understood in the light of the conflicting demands for forest biomass, by the three fundamental economies associated with forests:

1. Nature's economy of essential ecological processes.
2. The survival economy of basic needs satisfaction of the people.
3. The market economy of industrial-commercial demands.

Nature's economy of essential ecological processes generates a demand on the products of the forest in terms of the maintenance of the stability of soil systems and the hydrological balance of the forest ecosystems. In ecologically sensitive ecosystems like upland watersheds, this economy becomes the most crucial one and should be given the necessary attention in forest management. Neglecting this economy in upland watersheds will imply tremendous negative externalities to the national exchequer as relief for regular floods and drought, which are easily described as loss due to nature's fury. The survival economy of basic needs satisfaction reflects the requirement of forest biomass of the people living in and in the vicinity of forests in terms of fuelwood, fodder, fruits, nuts, green manure, small timber, etc. In forest areas where human settlements have either existed or are in the vicinity, the requirements of the survival economy have been satisfied all along without any major ecological damage. However, under certain situations, the pressure of the survival economy can be substantial and its neglect can lead to the unexpected and rapid degradation of forest resources.'

The market economy of industrial-commercial demands coastitutes the forest biomass demand of the total market system in the formal market economy. It includes the demand for pulpwood. plywood. furniture, house construction, etc., as well as fuelwood for the urban market. The demand for biomass from the urban industrial sector as well as the survival requirements of people have increased dramatically in the last century in India. There are many examples which illustrate that the growth of forest based industries is disproportionately beyond the ecological limit of the renewable productivity of nature. under the present system of management. The growth in population adds another significant demand for biomass for domestic purposes. It has, however, not been recognised, quantified and internalized in formal forest management. In the perspective of the increasing demand from survival and market economies, the biomass requirement for nature's economy are systematically sacrificed and nature's needs remain totally unfulfilled. In due course, this leads to the ecological destabilisation of the forest ecosystems.

Again, the conflict between survival and market economies assumes large proportions when most of the biomass produced is cornered by the economically powerful groups through the formal official mechanisms while the basic needs of fuel, fodder and small timber of the economically weak remain unsatisfied due to their weak political and economic status. Obviously, these three diverse economies push for the satisfaction of their demands either silently or loudly, resulting in both overt and covert conflicts over forest resources. The tacit and invisible nature of these conflicts combined with a mistaken or incomplete understanding of these diverse economies can result in the lack of perception of conflicts between them. The failure to perceive conflicts in forest use can only aggravate the forestry crisis. It is thus absolutely essential to understand and perceive the covert as well as explicit demands on forest resources to evolve a forest policy that is ecologically sensitive and socially just, so that our forest and land resources can be used for the overall satisfaction of the needs of the nation and the people in an equitable and sustainable manner.

The conceptual framework based on the three economies adopted here differs from the conventional categories related to forest resources. We do not use the dichotomy between quantity/quality as categories of forest use that Hallsworth, for example, has used. According to him, Human demands on the forest fall into two categories. First a demand for forest quantities; for the actual things a forest can produce: for timber, for food and for space for cultivation or for grazing; secondly, a demand for forest qualities; for the effects that forests have on the environment of man-to protect supplies of water, provide havens for wild-life and maintain the pool of genetic resources, to protect the soil against erosion, and to provide space for recreations. The dichotomy between quantity and quality, as between tangible and intangible, fails to capture the reality of those village women in Garhwal who launched the Chipko movement because for them water from the forest is a more significant product than wood and timber. For them water is not an 'intangible' produce or a mere quality. It is more tangible and basic in their sustenance economy than the commercial wood extracted and exported from their forests.

We have also avoided the categories of private versus public demands because they fail to show adequately that what are called individual or private needs at the local level are located in the public context of a village community, and what usually passes as public policy in forestry is a government policy which is totally subservient to commercial/industrial interests. The 'privatisation' inherent in a commercially-oriented forest policy is What has been challenged by movements like Chipko which demand a more effective social and public control over forest resources. The official management of forest resources in the name of 'national' or 'public' interest has, in reality, privatised people's 'common resources. This privatization trend which was rooted in the colonial period has survived to the present day. As Chhattarpati Singh observes, 'The class or "public" which more often than not benefits from such acquisitions is the rich. This is patently true in the acquisition of common land, especially forests. The public of the "public purpose" for which forests have been acquired constitutes all but the forest dwellers. We have therefore preferred to distinguish human demands on the basis of use for survival and use for commerce and industry, and have included nature's demands to avoid an anthropocentric bias as well as to have a context to locate forest movements like Chipko which struggle for forest conservation to respect nature's rights, and not just to assert their own rights to forest resources.

Bibliography

Anderson, J.T.: *Farm Forestry Practices for the Students of Vocational Agriculture*, Asiatic Pub, Delhi, 2008.

Anthony B.: *Applied Ethnobotany: People, wild Plant use, and Conservation*. London, Earthscan, 2001.

Babweteera, F.: *Influence of Gap-size and Age on the Diversity and Abundance of Climbers in Budongo Forest Reserve*, Uganda. M. Sc., Makerere University, 1998.

Bagyaraj, D. Joseph: *Microbial Biotechnology for Sustainable Agriculture Horticulture and Forestry*, New India Publishing Agency, Delhi, 2011.

Bahati, J.: *Impact of Arboricidal Treatments on the Natural Regeneration of Species Composition in Budongo Forest*, Uganda. M. Sc., Makerere University, 1995.

Bor, N. L.: *A Manual of Indian Forest Botany*, Asiatic Pub, Delhi, 2010.

Chakrabarty, Falguni: *Adaptation of the Santals to the Hill-Forest Environment*, APH, Delhi, 2012.

Connor Ogorzaly: *Economic Botany: Plants in our World*. Boston, McGraw-Hill, 2001.

Falguni, C: *Adaptation of the Santals to the Hill-Forest Environment*, APH, Delhi, 2012.

Jackie C. Wooten : *Ethnomedicine and Drug Discovery*, Amsterdam, New York, Elsevier, 2002.

Jerram, M.R.K.: *A Text-Book on Forest Management : In Agriculture, Horticulture and Forestry*, Asiatic Pub, Delhi, 2006.

Judith, S.: *The Natural History of Medicinal Plants*. Portland, OR, Timber Press, 2000.

Khare, C. P. *Encyclopedia of Indian Medicinal Plants: Rational Western Therapy, Ayurvedic and other Traditional Usage,* Berlin, New York, Springer, 2004.

Macself, A.J. : *Soils and Fertilizers*, Satish Serial Pub, Delhi, 2005.

Mani, A K; R Santhi and K M Sellamuthu: *Fundamentals of Forest Soils*, Satish Serial Pub, Delhi, 2008.

Miller, P. L.: *Some Dragonflies of the Budongo Forest*, Western Uganda. Opusc, 1993.

294 Handbook of Forest Botany Negi, S.S.: *Forest Soils*, International, Delhi, 2000.

Praveen, T.: *Advances in Forestry Research in India*, Cyber Tech Pub, Delhi, 2009.

Ramappa, B.T.: *Some Economic Aspects of India : Agriculture, Forestry and Industry*, Serials Publications, Delhi, 2011.

Reynolds, F.: *Chimpanzees of the Budongo Forest*, New York, Rinehart and Winston, 1965.

Reynolds, V.: *Budongo: A Forest and its Chimpanzees*, London, Methuen, 1965.

Sathe, T.V.: *A Textbook of Forest Entomology*, Daya, Delhi, 2009.

Sheelwant Patel: *Indian Forests : Soil Water and Bio-Environment Conservation*, Pointer, Delhi, 2005.

Shiel, D.: *The Ecology of Long-term Change in a Ugandan Rainforest*. D. Phil., Oxford University, 1997.

Sudharmai Devi,C.R. : *Analytical Procedures in Soil Science and Agricultural Chemistry*, Agrotech, Delhi, 2004.

Trivedi, P.C. : *Plant Physiology in Agriculture and Forestry*, Aavishkar Pub, Delhi, 2009.

Ward, H. Marshall: *Trees a Handbook of Forest-Botany*, Asiatic, Delhi, 2001.

Wickens, G. E.: *Economic Botany: Principles and Practices*. Dordrecht, Boston, Kluwer Academic, 2001.

Jerram. M.R.K. : *A Text-Book on Forest Management : In Agriculture, Horticulture and Forestry*, Asiatic Pub, Delhi, 2006.

Jha, L K: *Advances in Agroforestry*, APH, Delhi, 2009.

Kotwal, P.C. and Sujoy Banerjee : *Biodiversity Conservation in Managed Forests and Protected Areas*, Agrobios, Delhi, 2004

Kundu, S.S; O.P. Chaturvedi, J.C.Dagar and S.K. Sirohi : *Environment, Agroforestry and Livestock Management*,

International Book Distributing Co., Delhi, 2008.

Macself, A.J. : *Soils and Fertilizers*, Satish Serial Pub, Delhi, 2005.

Mohd. Azharul Haque and Tanmoy Rudra : *Biodiversity Extinction and Deforestation* , Jnanada Prakashan, Delhi, 2010.

Negi, S.S.: *Forest Soils*, International, Delhi, 2000.

Osmaston, A.E.: *A Forest Flora for Kumaon*, BSMPS, Delhi, 1994.

Prabhas Chandra Sinha : *Climate, Forest, Biodiversity and Desert*, SBS Pub, Delhi, 2006.

Puri, Sunil and Pankaj Panwar : *Agroforestry : Systems and Practices*, New India Pub, Delhi, 2007.

Reynolds, V.: *Budongo: A Forest and its Chimpanzees*, London, Methuen, 1965.

Index